Invertebrates of Streams and Rivers

A key to identification

Invertebrates of Streams and Rivers

A key to identification

Michael Quigley

Head of Studies in Environmental Biology
Nene College, Northampton

Edward Arnold

© Michael Quigley 1977

First published 1977
by Edward Arnold (Publishers) Ltd
41 Bedford Square, London WC1B 3DQ

Edward Arnold (Australia) Pty Ltd
80 Waverley Road, Caulfield East
Victoria 3145, Australia

Reprinted 1980, 1986

ISBN 0 7131 0091 5

All rights reserved. No part of this publication may be reproduced, stored in a retrieval system, or transmitted in any form or by any means, electronic, mechanical, photocopying, recording or otherwise, without the prior permission of Edward Arnold (Publishers) Ltd.

Set in 10 on 11 pt. Monotype Plantin Light
Printed in Great Britain by
Spottiswoode Ballantyne Ltd.,
Colchester and London

Foreword

My first introduction to the fascinating world of freshwater biology was a lecture by John Clegg at the Haslemere Educational Museum in Surrey. It was there under the incandescent glare of an ancient but efficient microprojector that I first saw some of the fantastic range of structure of the invertebrates which may be found in fresh flowing water. Since that day, I have never ceased to be amazed by the wealth of life that can be collected from a stream or river with nothing more sophisticated than a pond net and a white tin tray. One haul can provide sufficient material to keep a student happy for many hours, observing, probing and relating the adaptations to the ecology of the organisms.

The only real problem when embarking on such a study is one of identification, for knowledge, however detailed, is unsatisfactory if you haven't got a name to 'hang it on'.

Here is a nice book written by someone who not only knows his subject and is willing to share his enthusiasm for it with others but someone who understands the problems of a budding biologist faced with the bewildering variety of freshwater life.

It is a very welcome addition to the literature on flowing water and one which must help to 'open up' the habitat and its occupants to a much wider circle of admirers.

      David J. Bellamy
      Ph.D., B.Sc., F.L.S.
      Senior Lecturer in Botany
      Durham University

To David W. Roberts B.Sc., F.Z.S.
with gratitude

Preface

In this volume, I have provided an illustrated key to freshwater animals, for use in ecological studies, which I hope will be a valuable aid in the identification of invertebrates found in running water habitats.

It has been specifically designed to assist those who are likely to pursue some ecological work in their studies and would be particularly suited to beginners to fieldwork in 6th forms, and students on more advanced courses. There is no reason, however, why this guide should not be used by 5th form students in secondary schools, when perhaps for special study work, such a guide could prove useful.

The value of ecological studies in biology syllabuses in secondary schools, colleges and universities has been increasingly appreciated in recent years: fieldwork provides the biology student, or interested amateur naturalist, with a marvellous opportunity to make original discoveries, and provides him with an insight into the inter-relationship between living organisms and their environment.

Identification is the basis of all good fieldwork, yet there are comparatively few guides to the organisms of specific habitats available, and exhaustive searching through many books is often a daunting experience, especially to the beginner. It is hoped that this key will go some way to solving this problem, and will make identification an exciting challenge to the user.

This guide concentrates on the invertebrate animals likely to be found in the running water of streams and rivers, since a freshwater habitat is an interesting and rewarding start to any ecological work, and in most instances freshwater in some form is generally relatively accessible. Streams and rivers, with their wealth of invertebrate animals, variously adapted to survive in constantly flowing water provides a most stimulating prospect for the beginner.

As this is a book for beginners, the guide contains a selection of those invertebrate animals that are likely to be encountered in running water, and only the most common species are considered and named. In most cases, the various keys to aid in the identification will provide the user with sufficient information to identify a specimen collected to its nearest group.

Too often the beginner expects to be able to identify to species level but this is often an impossible task. In some cases it is comparatively easy, but where a rich diversity of species within a particular animal group is present the task may be formidable indeed. In realistic terms only the true specialist in a particular animal group can hope to identify all the species. The family Chironomidae (midges) may serve to illustrate the problem, and put the beginners mind at rest. The larval stages of this particular insect family are found in most aquatic habitats and contain close to 400 species. In this and other instances it is quite acceptable to identify such species either to the family group or perhaps to the genus level.

Michael Quigley
Nene College
Northampton

Contents

	page
Key to the major invertebrate groups or phyla found in streams and rivers	1
Porifera	2
Coelenterata	2
Platyhelminthes	3
Annelida	5
Oligochaeta	5
Hirudinea	7
Arthropoda	10
Crustacea	12
Isopoda	12
Amphipoda	13
Insecta—adults	13
Hemiptera	14
Coleoptera	18
Insecta—nymphs	22
Plecoptera	23
Ephemeroptera	33
Odonata	43
Insecta—larvae	48
Diptera	49
Trichoptera	52
Coleoptera	64
Insecta—pupae	65
Arachnida	66
Mollusca	67
Gastropoda	68
Prosobranchiata	69
Pulmonata	72
Lamellibranchiata	75
List of references for further reading	79
Key to the invertebrate animals of streams and rivers	80
Glossary	82

Acknowledgements

I am greatly indebted to the following publishers and authors for allowing me to redraw a number of illustrations from their books—The Freshwater Biological Association: Scientific Publication No. 13: *A Key to the British Fresh- and Brackish-Water Gastropods* by T. T. Macan; Scientific Publication No. 16: *A Revised Key to the British Water Bugs (Hemiptera-Heteroptera)*, by T. T. Macan; Scientific Publication No. 20: *A Key to the Nymphs of British Species of Ephemeroptera* by T. T. Macan; Scientific Publication No. 17: *A Key to the Adults and Nymphs of the British Stoneflies (Plecoptera)*, by H. B. N. Hynes; Liverpool University Press: *The Biology of Polluted Waters* by H. B. N. Hynes; Chapman and Hall Ltd., *Animal Life in Freshwater* by Helen Mellanby; Longman Group Ltd., *A Guide to Freshwater Invertebrate Animals* by T. T. Macan; Hutchinson Publishing Group Ltd., *Caddis Larvae* by Norman E. Hickin; Collins Publishers: *Dragonflies* by Philip S. Corbet, Cynthia Longfield and N. W. Moore.

My grateful thanks goes to my colleagues Pauline Wright, David George, Barry Alcock and Norman Copland at Nene College who have made valuable suggestions as to content and encouraged me during the preparation of this book; to students past and present who have used the guide in its variety of forms.

I wish to thank the publishers of this book, Edward Arnold Ltd., and in particular the Science Editor for advice and encouragement before and during publication.

Finally my most grateful thanks goes to my wife for her patience, faith and advice over the months of preparation. Without her encouragement the book would never have been completed.

If I have inadvertently forgotten to mention anyone, my apologies and thanks.

Cover photograph by Heather Angel, Biofotos

How to use the Guide

The object of the guide is to provide the user with sufficient information to assign an invertebrate animal collected to its correct classificatory groups.

In order to achieve this, information is given in terms of external characteristics, size, colour and occurrence of the more common invertebrates of running water.

Turn to the beginning of the guide—the key to the major animals groups, the PHYLA.

1. Against each number on the left-hand side of the page are two contrasting statements. Read these statements carefully and decide which of them most closely applies to the specimen being identified.
2. At the end of each statement is either a number or the name of the phylum to which the animal belongs.
3. If there is a number at the end of the statement it refers to the next pair of statements you should consider in order to determine the phylum to which your specimen belongs. This process continues until you reach the phylum name.
4. When the phylum has been determined turn to the appropriate section indicated by the page reference.
5. In each section, unless there are only a few animals considered, further keys are provided to assist you in your identification. The procedure for further identification is similar to that described above.
6. The following aids to successful identification are provided for you:
   (a) diagrams of a representative animal from each of the groups considered. These diagrams illustrate the parts of the animal which you should look at when using the keys,
   (b) clues on what to look for when examining your animal
   (c) when the name of a particular animal group is determined, a diagram or diagrams of a representative type is given. Each representative animal possesses the general characteristics of the particular animal group.
7. Finally, information is given about the common species within a particular group. When the species level is reached read the information about external characteristics, colour, size and where such species are likely to be found carefully. A detailed diagram of each animal is given, unless it is so like the preceding species. In such instances only that part which serves to distinguish them is drawn.

Symbols

The only symbols used in the guide are those which serve to denote the size of the animals considered. The size indicated is that of the maximum length, width or height attained, whichever is applicable.

├───────────────┤

This indicates the maximum length a particular animal is likely to attain. It should be appreciated that your specimen may be smaller than the size indicated.

This symbol is also used to indicate width of Gastropod mollusc shells and length of Lamellibranch mollusc shells.

├───────┼───────┤

This symbol indicates two lengths. The distance between the first two vertical bars indicates the maximum length of the animal whilst the distance between the first and third vertical bars indicates the maximum length of the case. Reference: Trichoptera Larvae.

I

This indicates the maximum height of the shell in Gastropod molluscs and the maximum width of the shells in Lamellibranch molluscs.

├───────────────┤ × 10

This indicates that the actual measurement given should be multiplied by 10 to obtain the maximum length or width of the particular shell.

Key to the major invertebrate groups or phyla found in streams and rivers

1 Body sponge-like, forming encrusting masses on submerged objects

☐ Phylum: Porifera (sponges)
see page 2
Body not forming encrusting masses and not sponge-like
☐ 2

2 Body small, tubular with a fringe of about 8 tentacles at the free end. Like small sea-anemones. Radially symmetrical. Attached by one end to plants etc. Body contracts when disturbed

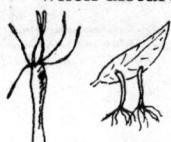

☐ Phylum: Coelenterata
(sac-like animals)
see pages 2–3
Unlike the previously mentioned animals
☐ 3

3 Body small, flattened dorso-ventrally, rather leaf-like in appearance. Body bilaterally symmetrical and unsegmented

☐ Phylum: Platyhelminthes
(flatworms)
see pages 3–5
Body bilaterally symmetrical but unlike the previously mentioned animals
☐ 4

4 Body elongated, worm-like and made up of a large number of segments. No obvious appendages (projections from the body surface)

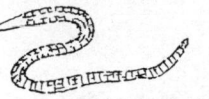

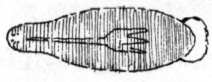

☐ Phylum: Annelida
(segmented worms)
see pages 5–10

Body either made up of many fewer segments and with obvious appendages, or body unsegmented without obvious appendages, but a shell or shells present
☐ 5

5 Few body segments, obvious jointed appendages usually present arising from some or all of the segments. Some may appear rather worm-like but with many fewer segments. Head well-developed. Some live in cases constructed of stones or vegetable material

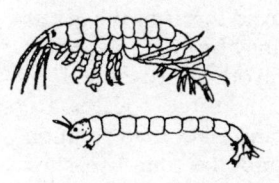

☐ Phylum: Arthropoda
(jointed-limbed animals)
see pages 10–66

1

Body soft, without obvious segments. Partially or almost totally covered by a hard, inflexible shell or shells
☐ Phylum: Mollusca
   (soft-bodied animals)
   see pages 67–78

Phylum: Porifera

This phylum is poorly represented in freshwater. The majority of sponges are found in marine habitats. In freshwater the phylum is represented by a member of the family Spongillidae.

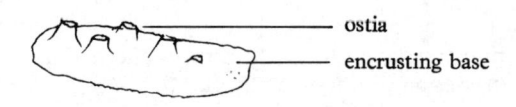

Family: Spongillidae

Description
Body flat and encrusting, pierced with varying sized small holes, ostia. The texture is very fine.
Colour: White or greenish
Spongilla fluviatilis

Occurrence
Encrusting masses on stones, plants, submerged roots etc.

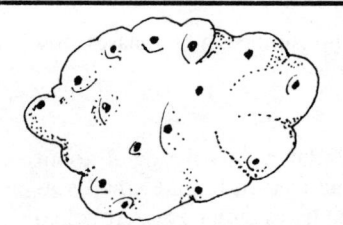

Phylum: Coelenterata

Like the previous phylum this one is also poorly represented in freshwater habitats. Those present are hydroid forms of the genus *Hydra*. Three species may be found but apart from the green form the others are difficult to identify accurately. The three species are more typical of ponds and lakes than running water, although they may be found in slow running water where there is much vegetation.

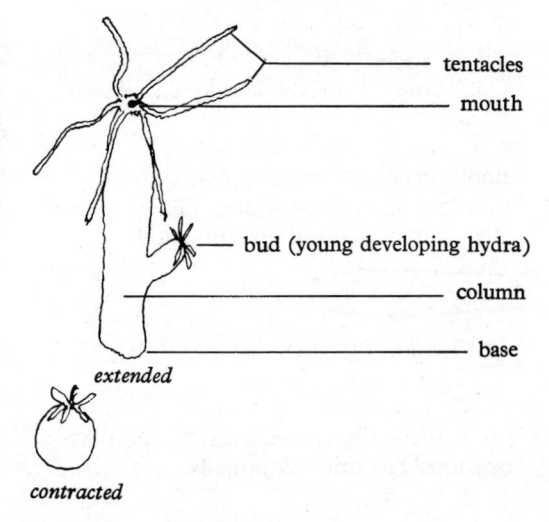

extended

contracted

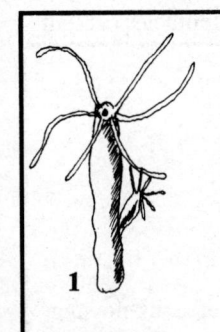

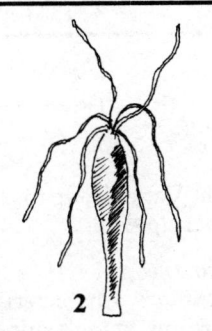

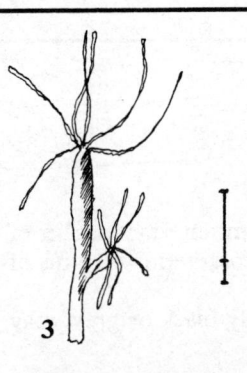

Description
1 Colour: Green
 Hydra viridis
2 Colour: Brown
 Basal part of body usually slender when animal is extended
 Hydra oligactis

3 *Colour: Greyish-brown*
 Whole body slender when extended
 Hydra attenuata
 Length 2–12 mm approx.

Occurrence
Attached to vegetation in ponds, lakes and slow-running streams

Phylum: Platyhelminthes

Some of the species of the order Tricladida are characteristic of running water whilst others are more characteristic of standing water. However, none of them are restricted to one or other of these habitats. The species dealt with here are generally more characteristic of running water.

All the species considered here belong to the family Planariidae.

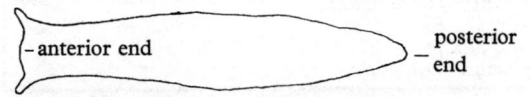

The simplest way of identifying flatworms is to look closely at the shape of the head, and determine whether tentacles are present or absent.

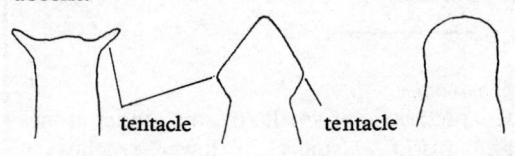

tentacles present   *tentacles absent*

The number of eyes or eye-spots and general colouring of the body will also help in identifying flatworms.

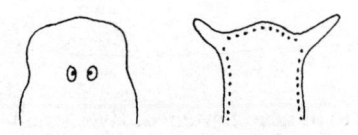

1 Two eyes, tentacles present
  ☐ *Crenobia alpina* and *Dugesia tigrina*
     see page 4

2 Two eyes, no obvious tentacles
  ☐ *Phagocata vitta*
     see page 4

3 Many eye-spots, tentacles present
  ☐ *Polycelis felina*
     see page 4

4 Many eye-posts, no obvious tentacles
  ☐ *Polycelis nigra*
     see page 5

1 Two eyes, tentacles present

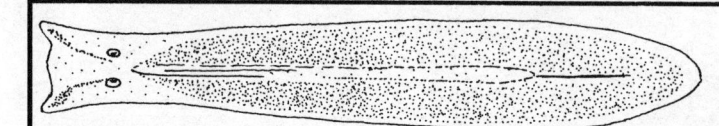

Crenobia alpina

Description
Eyes some distance from anterior margin of head. Tentacles on latero-anterior margins of head.
Colour: Variable, usually black or grey, may be brown.

Length: Variable, up to 20 mm
Crenobia alpina

Occurrence
Under stones, Characteristic of swift-flowing, cool streams, at high altitudes.

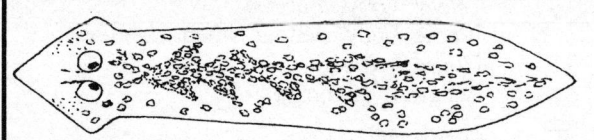

Dugesia tigrina

Description
Usually has two eyes some distance from anterior margin of the head, occasionally more than two eyes. Head triangular. Tentacles lateral.
Colour: Mottled grey-brown

Length: Up to 25 mm
Dugesia tigrina

Occurrence
In the main restricted to the southern parts of Britain. Found in the slower reaches of streams and rivers.

2 Two eyes, no obvious tentacles

Phagocata vitta

Description
Anterior margin of head flattened. Eyes small and close together.
Colour: White
Length: Up to 12 mm
Phagocata vitta

Occurrence
Characteristic of streams at quite high altitudes. Many occur at lower altitudes in head waters of streams.

3 Many eye-spots, tentacles present

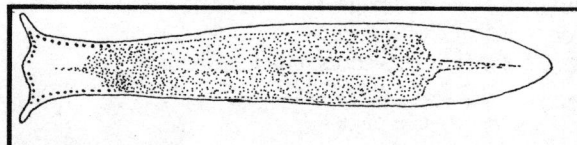

Polycelis felina

Description
Eye-spots situated on lateral margins of the head, tentacles variable in length.
Colour: Black or brown-black on dorsal surface, paler on ventral surface.
Length: When extended up to 20 mm
Polycelis felina

Occurrence
Characteristic of small streams, under stones and plants. Usually in lower reaches of streams where temperatures fluctuate.

4 Many eye-spots, no obvious tentacles

Polycelis nigra
├────┤

Description
Body broadest at front part. Eye-spots on lateral and anterior margins of the head.
Colour: Usually black, sometimes brown.

Length: Variable, up to 12 mm
Polycelis nigra

Occurrence
Characteristic of lowland streams, ponds and lakes.

Phylum: Annelida

The classes of this phylum which have representatives in freshwater habitats are the Oligochaeta and the Hirudinea (leeches).
Key to classes:

1 Body long and rather thread-like, cylindrical, usually divided into a large number of segments. No obvious head. Bristles or chaetae protrude from the surface of the segments (microscopic examination)

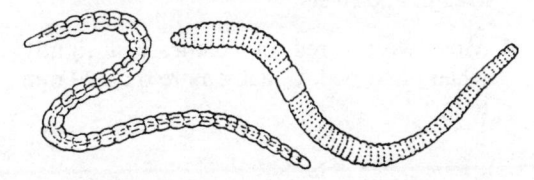

☐ Class: Oligochaeta
   see pages 5–7

2 Segmented worms but body more flattened. Bristles are absent from the body segments. A sucker is present at each end of the body. Typically attached to stones, may be difficult to remove

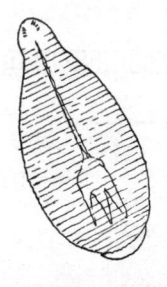

☐ Class: Hirudinea
   see pages 7–10

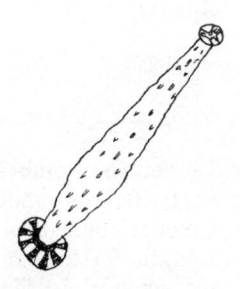

Class: Oligochaeta (Fresh-water worms with bristles or chaetae)

A number of families are represented in freshwater. These are usually quite easy to identify.

Identification beyond families is difficult and requires careful microscopic examination. For this reason no details of genera or species are given.

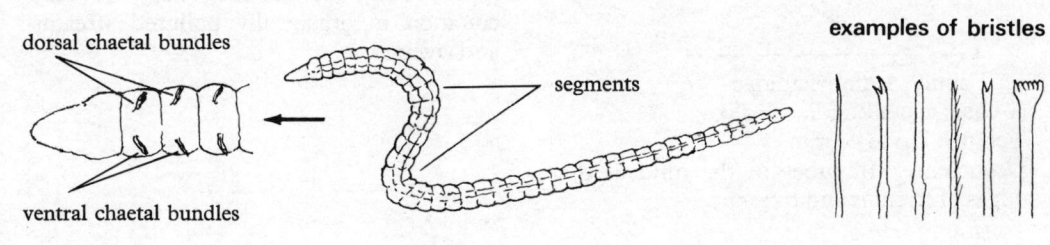

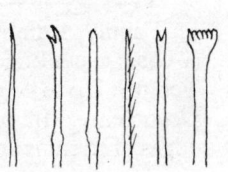

dorsal chaetal bundles

ventral chaetal bundles

segments

examples of bristles

Examine carefully and determine
(a) length and shape
(b) colour
(c) if necessary, the types of and the number of chaetae in the dorsal and ventral bundles. These should be observed with a good microscope if it is felt necessary. Details of chaetae given in brackets

Key to families:

1 Worms usually longer than 30 mm in length
☐ 2

Worms up to or shorter than 30 mm in length
☐ 4

2 Like small earthworms with a distinct clitellum
Head slightly pointed, posterior region ends squarely

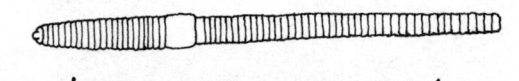

☐ Family: Lumbricidae
(2 chaetae per bundle, all alike)
Colour: Yellowish-brown, pink
Length: Up to 50 mm
Occurrence: In the substratum of mountain streams or in the algae or mosses covering stones

Worms long and thin
☐ 3

3 Worms up to 40 mm long, may be longer. Bright red in colour.

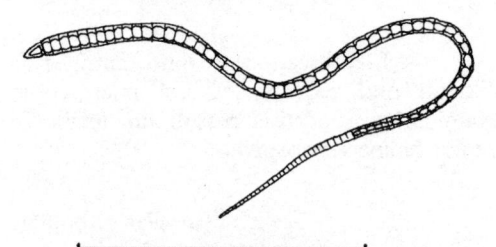

☐ Family: Lumbriculidae
(2 chaetae per bundle, all alike)
Length: Up to 40 mm
Occurrence: In tubes in the mud at the edges of streams and rivers

Worms very long and thin and thread-like. Up to about 500 segments.

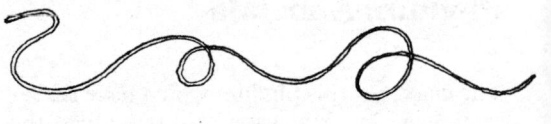

⊢──────────⊣ x 10

☐ Family: Haplotaxidae
(Dorsal and ventral bundles, chaetae simple pointed, one per bundle. Ventral chaetae longer than dorsal chaetae)
Length: Up to 300 mm
Occurrence: In the mud, near the banks of streams and rivers

4 Most worms red in colour, coil tightly when disturbed. Usually more than 20 mm long

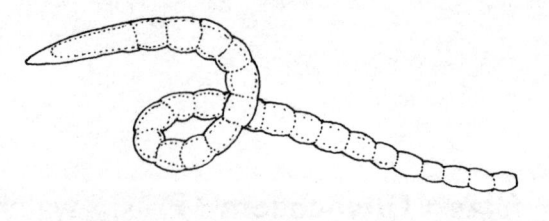

☐ Family: Tubificidae
(More than 2 chaetae per bundle, chaetae of various types)
Length: Up to 30 mm
Occurrence: Common in the mud of all types of aquatic habitats. May be very common in organically polluted streams and rivers.

Worms white or pinkish, from 2–20 mm long. May be found in chains of individuals

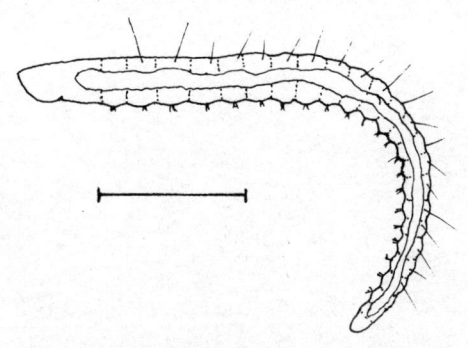

☐ Family: Naididae
(Hair chaetae may be absent from dorsal bundles, cleft chaetae in ventral bundles)
Length: Up to 20 mm
Occurrence: Mainly found in the mud of ponds and lakes but some, particularly members of the genus *Nais*, inhabit the mud of streams and rivers.

Class: Hirudinea (leeches)

The leeches presented in this guide are those that are typically found in streams and rivers. They are external parasites and feed on the blood and body fluids of other aquatic animals.

They differ from other annelids by possessing suckers, one at each end of the body, and by the general dorso-ventral (top to bottom) flattening of the body. They are usually found attached to stones or other objects in the water.

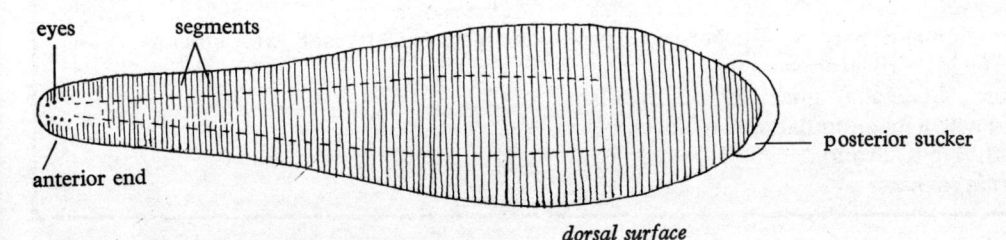

dorsal surface

Examine the worms and carefully determine:
(a) the number and position of the eyes on the head
(b) the prominence of the suckers
(c) the number of segments

Key to families:

1 Two pairs of eyes. Anterior and posterior suckers distinct from the rest of the body. Segments small and numerous (200+)

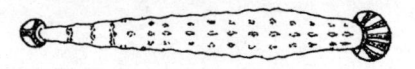

☐ Family: Piscicolidae
see page 8

More than two pairs of eyes
☐ 2

2 Usually three pairs of eyes, first pair may be small or unpigmented. Anterior sucker not very distinct from the rest of the body. Usually up to about 80 segments

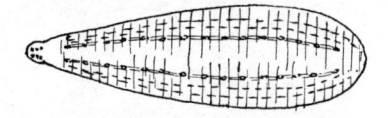

☐ Family: Glossiphonidae
see page 9

3 Four pairs of eyes in two transverse rows. Suckers indistinct from the rest of the body. Mouth large. Segments all about the same length

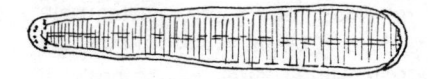

☐ Family: Erpobdellidae
see page 10

Family: Piscicolidae

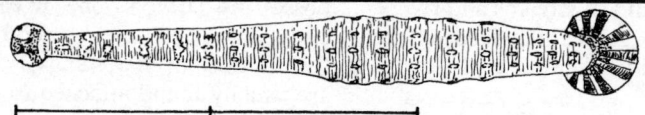

Piscicola geometra

Description
Body long and narrow. Suckers are prominent with a frilled appearance.
Colour: Generally greenish or reddish-brown with 8 longitudinal rows of spots.
Length: Up to 50 mm
Piscicola geometra

Occurrence
Characteristic of fast flowing streams

Family: Glossiphonidae

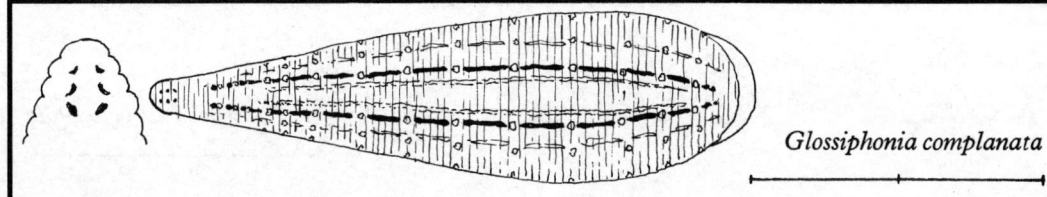

Glossiphonia complanata

Description
Elongated and somewhat oval. Head narrow, usually three pairs of eyes, first pair small and sometimes absent.
Colour: Greenish-brown with 6 longitudinal rows of cream-brown spots.
Length: Up to 40 mm
Glossiphonia complanata

Occurrence
In running water

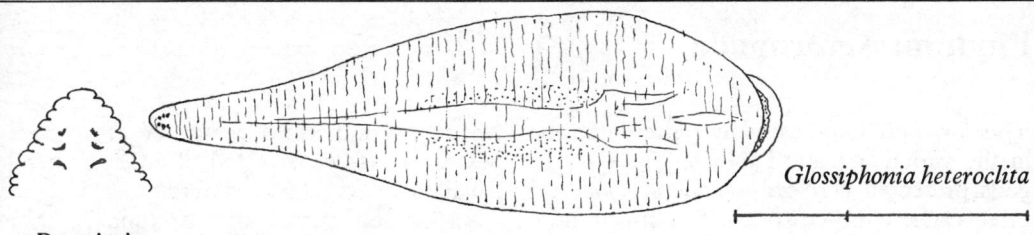

Glossiphonia heteroclita

Description
Body of similar shape to that of *G. complanata*. Three pairs of eyes, first pair may be unpigmented.
Colour: Yellowish with small black-brown markings
Length: Up to 40 mm
Glossiphonia heteroclita

Occurrence
Generally found in slow-running water

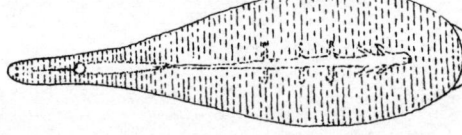

Helobdella stagnalis

Description
Body similar in shape to the previous two species. Head very narrow. One pair of eyes set close together.
Colour: Pinkish or greyish-green with surface flecked with black
Length: Up to 20 mm
Helobdella stagnalis

Occurrence
Slow running water where conditions are rather pond-like

Family: Erpobdellidae

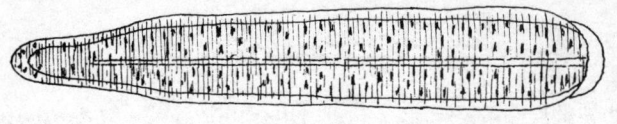

Erpobdella octoculata

Description
Body very elongated. Possesses four pairs of eyes.
Colour: Usually dark brown, sometimes brownish-red or even yellow
Length: Up to 40 mm when at rest, longer when extended
Erpobdella octoculata

Occurrence
Occurs in all kinds of freshwater habitats, can withstand moderate pollution in streams and rivers

Phylum: Arthropoda

This large phylum has many representatives in the various aquatic habitats. The arthropods presented here are members of one of three classes, the crustacea, the insecta and the arachnida.
Examine carefully and determine:

(a) whether the body is flattened or not
(b) whether the body is divided into distinct regions—head, thorax and abdomen, or not
(c) if distinct walking legs are present, and if so, the numbers

Note: An important group of arthropods possess a case constructed of either mineral particles or vegetable material, or live in specially constructed nets or tunnels. The above characteristics mentioned under (a) (b) and (c) may therefore be difficult to determine. However, when active in water such animals do protrude their heads and legs. For a more detailed examination the cased forms should be removed from their cases. Do not confuse with insect pupae which are largely inactive.

Key to classes:

1 Body typically flattened either dorso-ventrally (from top to bottom) or laterally (from side to side). Typically one pair of appendages for each body segment

☐ Class: Crustacea
see pages 12–13

Body may be dorso-ventrally flattened in some but not with one pair of appendages for each body segment
☐ 2

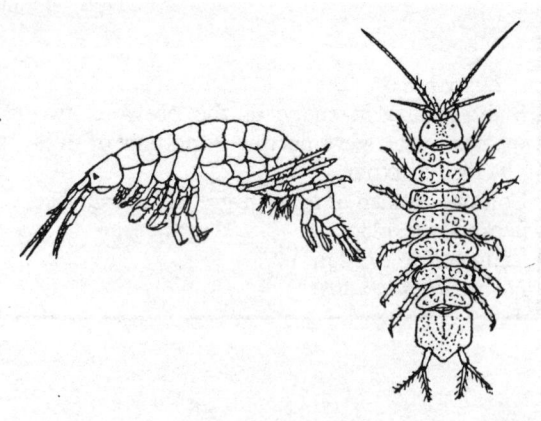

2 Body divided into three distinct regions, insect-like, with three pairs of legs
☐ 3

Body not divided into three distinct regions, legs may or may not be obvious
☐ 4

3 Surface of body appears hard and often shiny. Wings present, fore-wings (top) may be modified to form wing-cases (elytra)

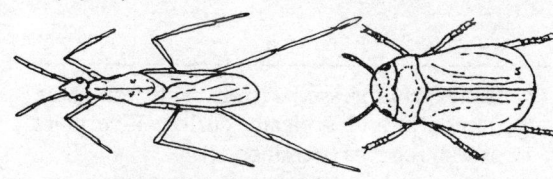

☐ Class: Insecta (adults)
see pages 13–21

Surface of body may not appear hard. Wings, if present, are small and functionless (wing buds). Abdomen terminates in two or three projections

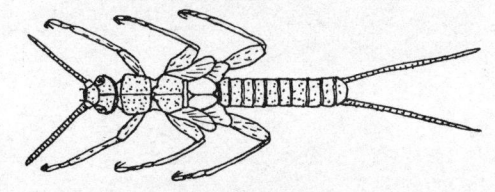

☐ Class: Insecta (nymphs)
see pages 22–47

4 Body may be rather maggot or worm-like. May have no obvious legs, fleshy projections from the segments or three distinct pairs of legs arising from the thoracic region. Some live in cases of vegetable material or mineral particles; some construct nets or tunnels.

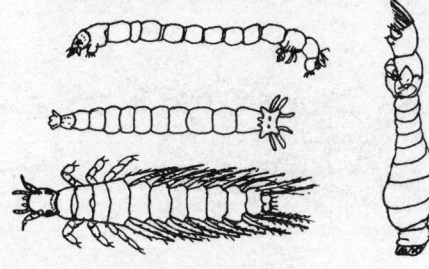

☐ Class: Insecta (larvae)
see pages 48–64

May be like insect larvae but encased in a protective cocoon, this may be made of body secretions which become hardened, small stones plugged with vegetable material. Some forms may be quite active, but the majority are inactive and are either buried in the substratum or attached to objects in the water

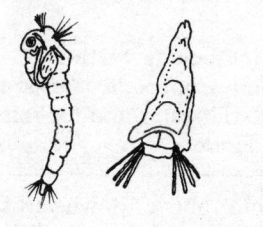

☐ Class: Insecta (pupae)
see page 65

Body rather globular in shape, with four pairs of obvious legs

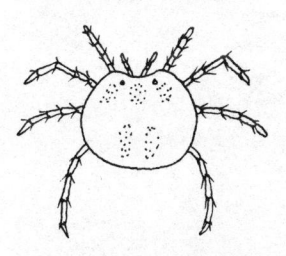

☐ Class: Arachnida (water mites)
see page 66

11

Class: Crustacea

This class is fairly well represented in freshwater habitats, the majority being found in still water of ponds and lakes. Many of the representatives are rather small and accurate identification requires microscopic examination.

    The smaller forms belong to the subclasses Phyllopoda and Copepoda and are more characteristic of still water and for this reason they are ignored here.

Examine carefully and determine:
(a) whether body is flattened dorso-ventrally or laterally
(b) the number of pairs of walking legs
Only two orders of crustaceans are dealt with here

---

1 Body flattened from above downwards (dorso-ventrally). Resembling woodlice. Large outer antennae. Seven pairs of walking legs (pereiopods) all similar

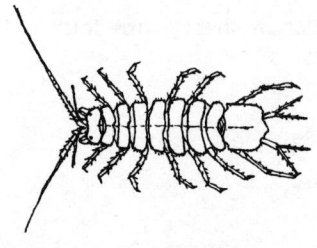

☐ Order: Isopoda
see page 12

2 Body compressed from side to side (laterally) and typically curled. Five pairs of walking legs (pereiopods)

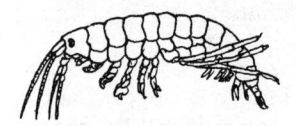

☐ Order: Amphipoda
see page 13

Order: Isopoda

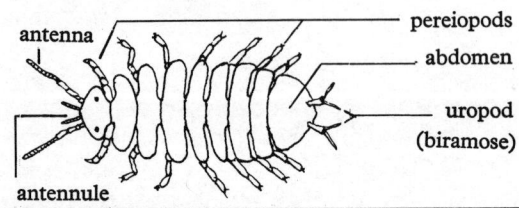

labelled diagram: antenna, pereiopods, abdomen, uropod (biramose), antennule

Description
Head small, thorax clearly segmented. Antennae at least three times as long as antennules. Abdomen with six pairs of appendages, the last pair biramose (uropods).
Colour: Transparent grey-brown
Length: Up to 25 mm
Asellus aquaticus

Occurrence
In ponds, lakes and slow-running streams. Particularly abundant in streams recovering from organic pollution

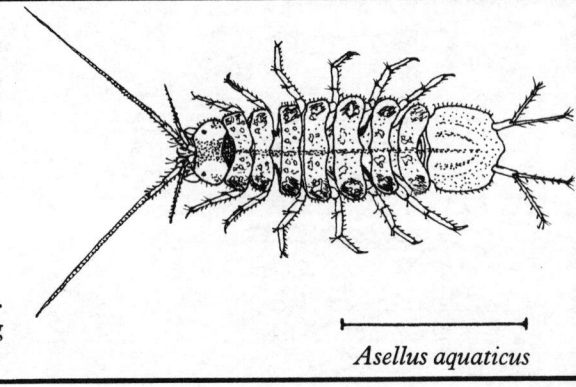

Asellus aquaticus

Order: Amphipoda

One species of the family Gammaridae is common in freshwater.

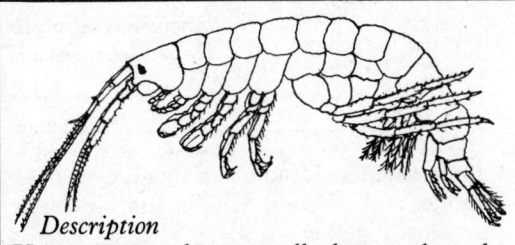

Gammarus pulex

Description
Upper antennules generally longer than the lower antennae. Second and third thoracic appendages are modified for grasping.
Colour: Pale reddish-brown to grey-brown
Length: Up to 30 mm
Gammarus pulex

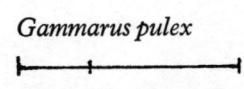

Occurrence
Characteristic of stony streams where they normally live under stones. In faster flowing streams generally found towards the sides of the stream.

Class: Insecta—Adults

Two insect orders are represented here, the Hemiptera (water bugs) and the Coleoptera (beetles).

1 Head prolonged into a beak-like structure (rostrum) in most species. Some have a blunt triangular head. Rostrum may be tucked under the head and only visible from the side or from beneath. Fore-wings are modified into wing-cases or elytra which are clearly divided into areas

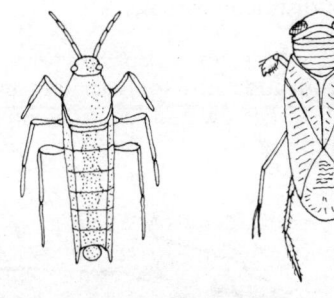

☐ Order: Hemiptera
see pages 14–17

2 Body generally oval in shape, fore-wings modified as elytra. Elytra hard and often sculptured, ridged and pitted. Hind wings membranous, usually folded away, may be absent. Prothorax usually large, Biting mouthparts

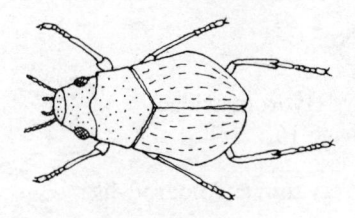

☐ Order: Coleoptera
see pages 18–21

1 Order: Hemiptera

Examine carefully and determine
(a) whether water or surface dwellers
(b) shape of body
(c) position and details of legs

Key to families:

1 Insects living in the water. Body rather shield-shaped, and flattened dorso-ventrally (from top to bottom). Head triangular with a beak. Rostrum concealed. Forelegs short, others long

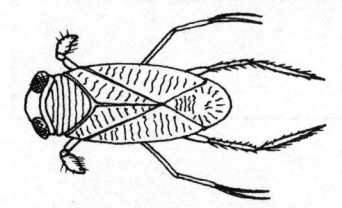

☐ Family: Corixidae
  see pages 14–16

Insects surface-dwellers. Body not shield-shaped, not distinctly flattened
☐ 2

2 Body very thin and rod-like. Legs long and stilt-like (Head narrow and long, antennae longer than the head)

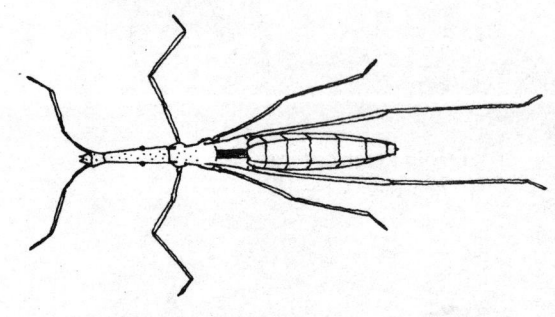

☐ Family: Hydrometridae
  see page 16

Body not very thin and not rod-like
☐ 3

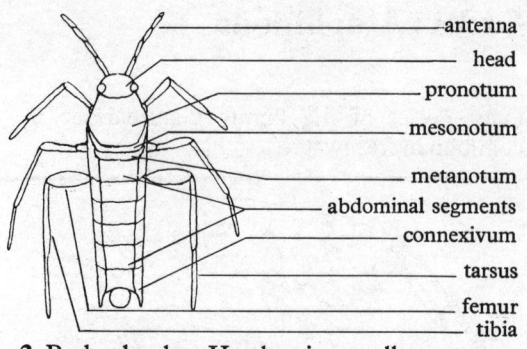

(Diagram labels: antenna, head, pronotum, mesonotum, metanotum, abdominal segments, connexivum, tarsus, femur, tibia)

3 Body slender. Head quite small, eyes near thorax end of head. Middle and hind-legs set close together

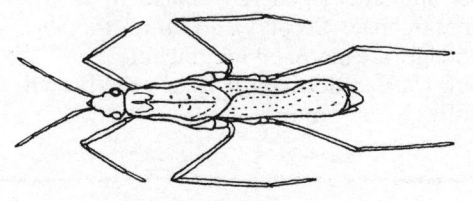

☐ Family: Gerridae
  see page 17

More stoutly built than the Gerridae. Middle legs roughly mid-way between the fore- and hind-legs. Middle and hind-legs not elongated

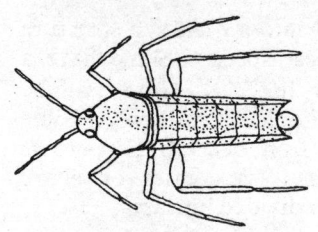

☐ Family: Veliidae
  see page 17

Family: Corixidae (lesser water boatmen)
These water bugs are common in most aquatic habitats, but only a few species are common in rivers and streams. Those that are found are more typical of regions where the water flow is slow.
(**Note:** In the nymph stage the abdominal segments can be seen.)
Look closely and determine:
  (a) size
  (b) whether the scutellum is visible or not
  (c) the number of segments of the antennae

1 Small, 3 mm or less in length, scutellum visible, antennae with three segments

☐ Sub-family: Micronectinae

2 Body longer than 3 mm, scutellum not visible and antennae with four segments

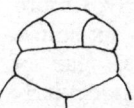

☐ Sub-family: Corixinae
see pages 15–16

1 Sub-family: Micronectinae

Description
Pronotum with a tubercle (hump) in the middle. Head with a dark central line.
Colour: Corium a pale ground colour with large, dark brown markings.
Length: Up to 3 mm
Micronecta poweri

Occurrence
Clean river and stream shallows with sandy bottoms and a slow flow

Micronecta poweri

Sub-family: Corixinae
(Examples)

Look closely at the following
(a) the legs
(b) the pronotum
(c) the patterns on the corium

Description
Pronotum and corium smooth and shiny. Inner end of tibia of second pair of legs tapering slightly at the point of articulation with the femur.

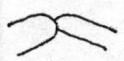

Claws on the second pair of legs shorter than the tarsi.
Colour: Dark and shiny
Length: Up to 13 mm
Corixa punctata

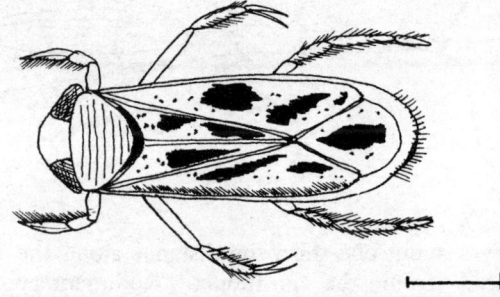

Occurrence
In river backwaters where the current is slow

Corixa punctata

Description
Tarsi on the third pair of legs with a dark, square mark at outer end.

Hind margins of femora of second pair of legs with few hairs. Regular markings on the corium.
Length: Up to 8 mm
Callicorixa praeusta

Callicorixa praeusta

Occurrence
Stagnant areas of streams and rivers, particularly regions which are organically polluted

Description
Body broad and rounded posteriorly. Pronotum with 6 pale stripes. Three to six spines on upper surface of femora of last pair of legs. Corium finely wrinkled and regularly marked transversely.
Colour: Dark brown–black
Length: Up to 10 mm
Sigara dorsalis

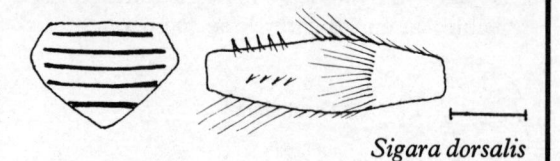

Sigara dorsalis

Occurrence
Habitats with clean, flowing water. Has a considerable range

Description
Similar to *Sigara dorsalis* but angles at the corner of the pronotum acute. More than 6 stripes on the pronotum. Seven to twelve spines on femora of last pair of legs.
Colour: Dark brown–black
Length: Up to 9 mm
Sigara falleni

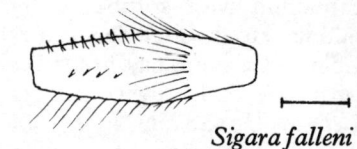

Sigara falleni

Occurrence
Lowland areas in slow flowing streams, ditches and rivers where conditions are alkaline, particularly in the south and west

Family: Hydrometridae

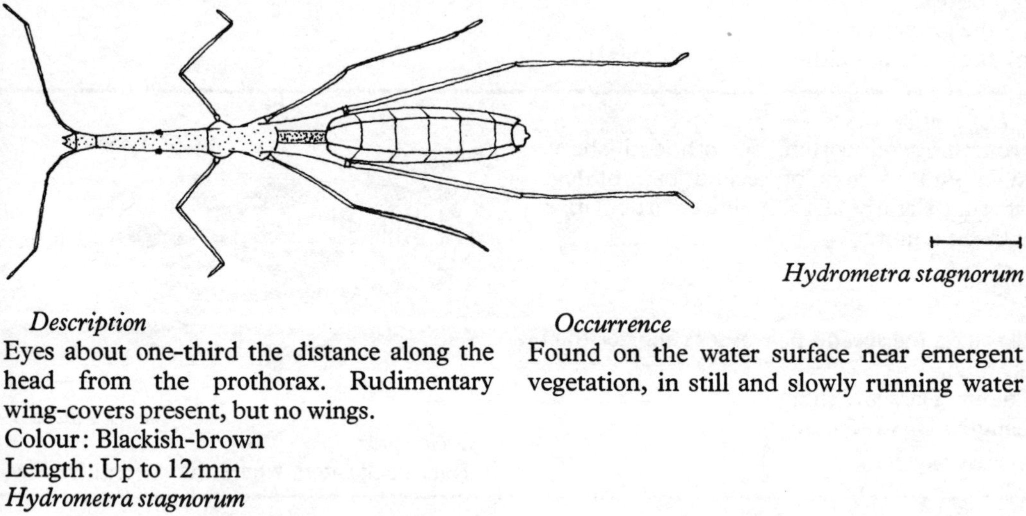

Hydrometra stagnorum

Description
Eyes about one-third the distance along the head from the prothorax. Rudimentary wing-covers present, but no wings.
Colour: Blackish-brown
Length: Up to 12 mm
Hydrometra stagnorum

Occurrence
Found on the water surface near emergent vegetation, in still and slowly running water

Family: Gerridae
(Example)

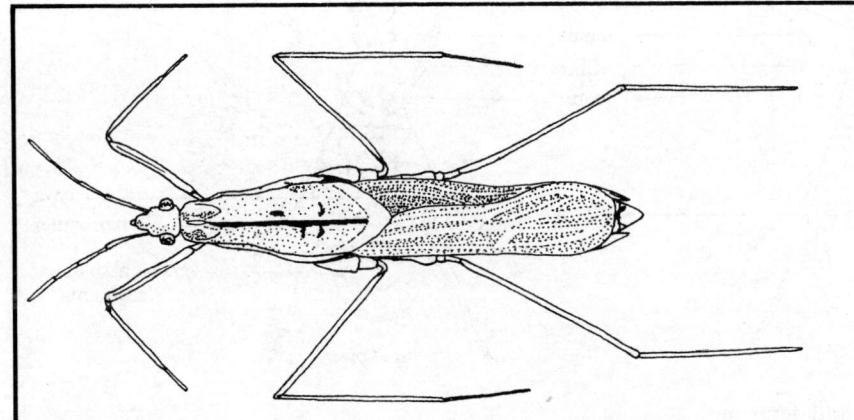

Gerris najus

Description
Antennae less than half body length. Middle femora as long or longer than hind leg femora. Sixth abdominal segment with two marginal spines.
Colour: Black-brown
Length: Up to 17 mm
Gerris najus

Occurrence
Occurs on margins of lakes, large streams and rivers

Family: Veliidae
(Example)

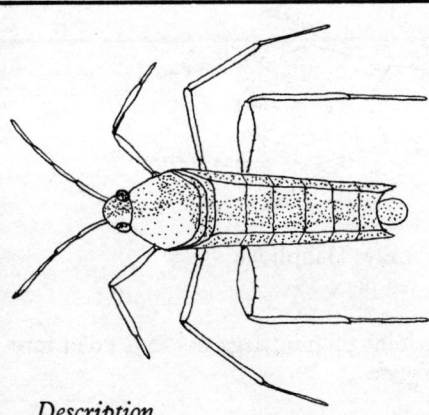

Velia caprai

Description
Body rather stout. Usually wingless. First abdominal tergite round, projects above connexivium. Edges of connexiva curved upwards.
Colour: Brown, with two orange lines down the back. Ventral surface of abdomen orange
Length: Up to 8 mm
Velia caprai

Occurrence
Usually occurs in running water, near banks in slower reaches

Order: Coleoptera

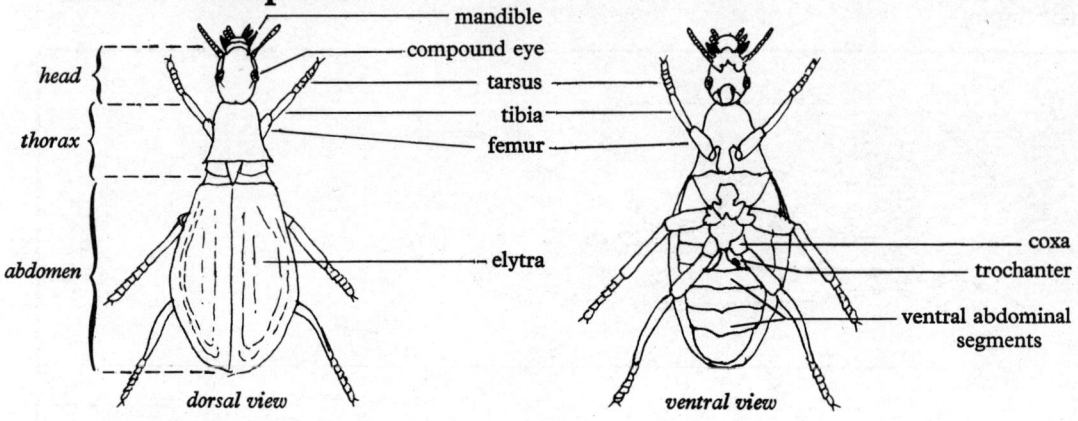

dorsal view — mandible, compound eye, tarsus, tibia, femur, elytra; head, thorax, abdomen

ventral view — coxa, trochanter, ventral abdominal segments

Examine carefully and determine:
(a) whether water or surface dwellers
(b) shape and details of legs
(c) size

Key to families:

1 Surface dwellers. Middle and hind legs short and broad, paddle-like. Eyes divided into two parts, lower part on underside of head

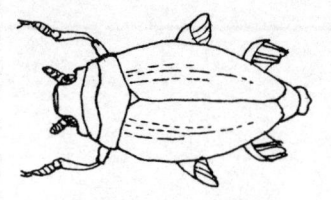

☐ Family: Gyrinidae
see page 19

Water dwellers. Middle and hind-legs not short and broad
☐ 2

2 First joint of hind-legs expanded into a large plate (examine from ventral surface) hiding leg attachment and much of abdomen. Antennae thread-like, 10-jointed. Usually less than 6 mm long

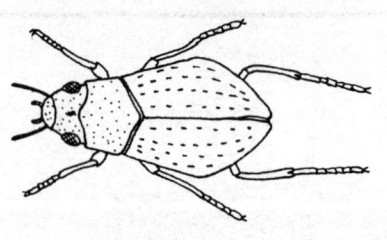

☐ Family: Haliplidae
see page 19

First joint of hind-legs not expanded into large plate
☐ 3

3 Hind-legs flattened, fringed with swimming hairs. Antennae thread-like, 11-jointed. Mandibles strong. Up to 15 mm long

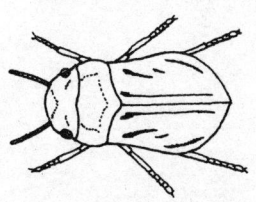

☐ Family: Dytiscidae
see pages 20–21

Hind-legs not flattened, last segment of tarsi long and bear a pair of long, sharp claws. Head can retract into thorax. Very small, up to 3 mm

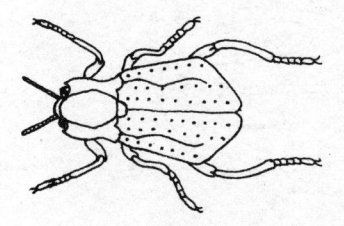

☐ Family: Helmidae (Elmidae)
see page 21

Family: Gyrinidae
(Example)

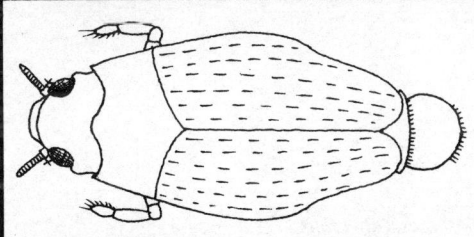

Gyrinus natator

Description
Small oval beetles, shiny black in colour. Legs yellow.
Length: Up to 10 mm
e.g. *Gyrinus natator*

Occurrence
Slow-running streams or stagnant water

Family: Haliplidae
(Example)

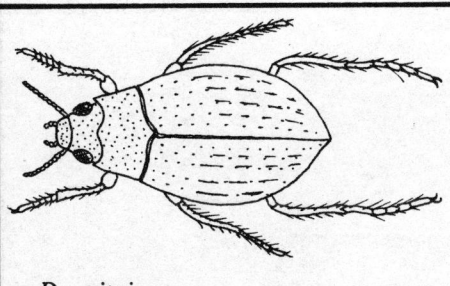

Haliplus fulvus

Description
Feeble swimmers. Yellowish or reddish (rusty) in colour with darker markings on the elytra.
Length: Up to 6 mm
e.g. *Haliplus fulvus*

Occurrence
Slow-running streams or stagnant water

Family: Dytiscidae
(Examples)

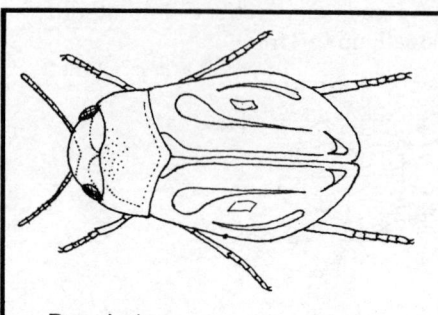

Platambus maculatus

Description
Reddish-yellow with strong black markings.
Length: Up to 8 mm
Platambus maculatus

Occurrence
Generally found in streams

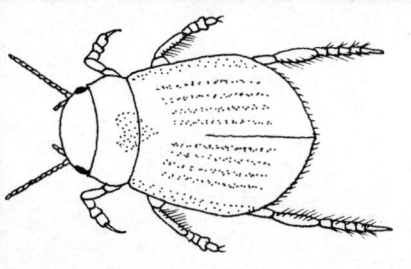

Derenectes depressus

Description
Thickly covered with short hairs. Head and thorax yellowish. Fore part of thorax has a narrow dark margin and the hind margin has two dark patches. Elytra yellowish with dark margins.
Length: Up to 5 mm
Derenectes depressus

Occurrence
Found in submerged vegetation in running water

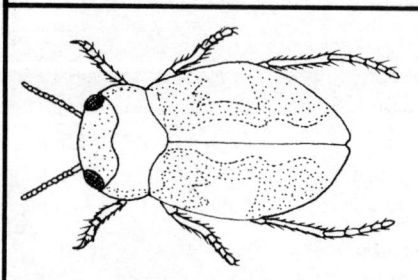

Hydroporus sp

Description
Very similar to members of the genus *Derenectes*. Very small and very variable in colour.
Length: Up to 5 mm
e.g. Genus: *Hydroporus*

Occurrence
Found in the lower reaches of streams

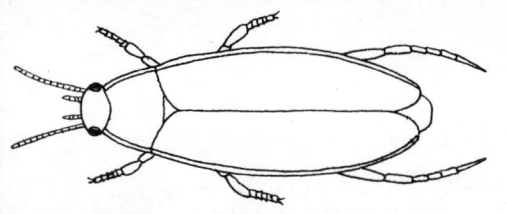

Ilybius fuluginosus

Description
Body long and narrower than other members of the family Dytiscidae. Claws of hind tarsi unequal. Tip of abdomen shows beyond elytra.
Length: Up to 15 mm
Colour: Bronze with yellow margins to elytra
Ilybius fuluginosus

Occurrence
Found in the lower reaches of streams and rivers. A mud-dweller

Family: Helmidae (Elmidae)
(Example)

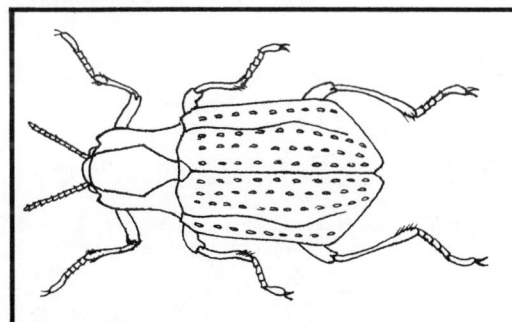

Helmis maugei

Description
Last joint of legs long, bearing long sharp claws.
Colour: Black, dark metallic bronze on elytra. Antennae brown with red bases. Tarsi reddish.
Length: Up to 3 mm
Helmis maugei

Occurrence
Fast flowing streams on stones, attached to vegetation

Class: Insecta
(nymphs)

The various orders of insects may be divided into two groups depending on the type of metamorphosis they undergo. The term hemimetabolous is used for those insects, considered here, which go through a simple metamorphosis. There is no pupal stage and in many cases the young resemble the adults in general form.

Examine carefully and determine:

(a) the number of tail appendages
(b) the presence or absence of gills
(c) the presence of an obvious protrusible labium

Key to orders:

1 Nymphs with two tails or cerci extending from the last abdominal segment. Abdominal gills absent. Legs end in two claws. Wing buds (pads) obvious in mature stages

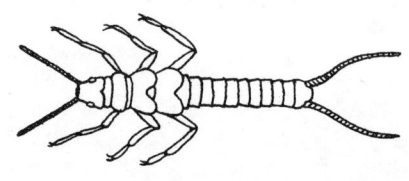

☐ Order: Plecoptera (stone flies)
see pages 23–32

Nymphs with three tails extending from the last abdominal segment
☐ 2

2 Tail appendages (2 cerci and 1 caudal appendage) long and thin. Gills present on abdominal segments. Legs end in one claw. Wing buds (pads) obvious in mature stages

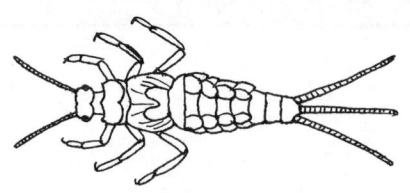

☐ Order: Ephemeroptera (may flies)
see pages 33–43

Tail appendages broad and flattened or triangular, may be very short. Gills absent. Labium modified into a protrusible 'mask'

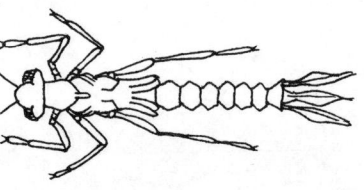

☐ Order: Odonata (dragonflies)
see pages 43–47

Order: Plecoptera

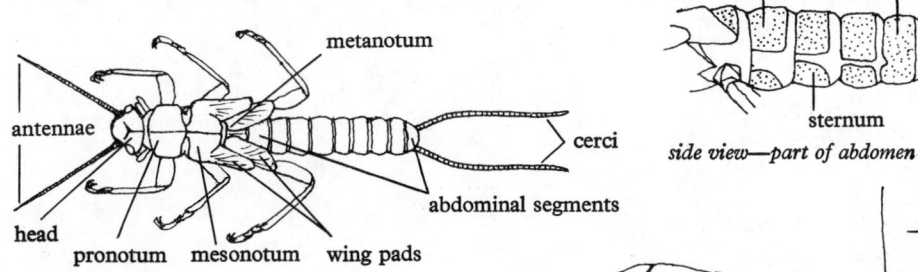

side view—part of abdomen (labels: tergum, complete abdominal ring, sternum)

(leg labels: femur, tibia, tarsus, claws)

(body labels: antennae, head, pronotum, mesonotum, wing pads, metanotum, abdominal segments, cerci)

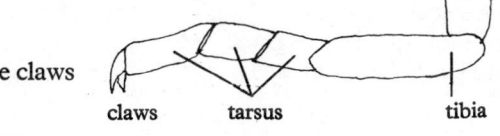

Examine the three segments next to the claws of the hind-leg—the tarsus region.

1 Each segment longer than preceding segments

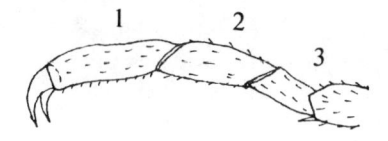

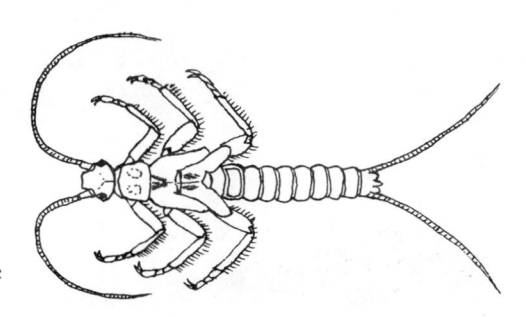

☐ Family: Taeniopterygidae
see page 25

2 The second segment is shorter than the first and third segments:

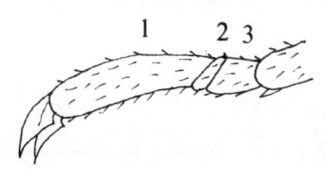

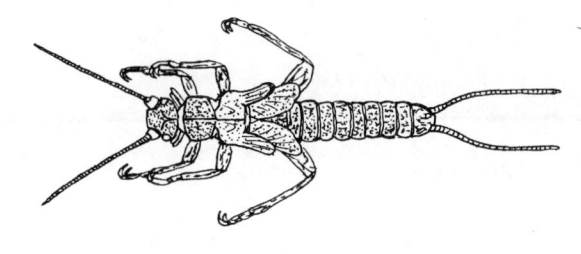

(a) Nymphs stout, wing pads set obliquely to the body. Hind-legs when stretched backwards greatly extend beyond the tip of the abdomen

☐ Family: Nemouridae
see pages 26–28

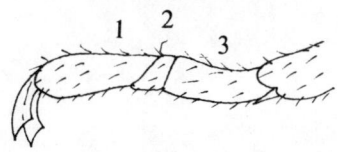

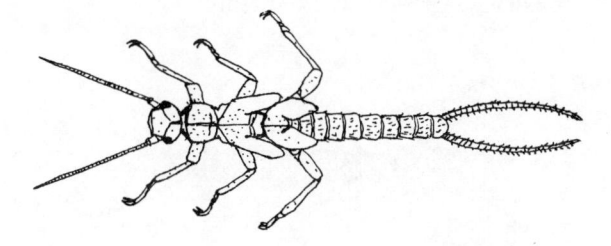

(b) Nymphs elongated and body cylindrical. Stretched hind-legs do not extend beyond tip of abdomen. First four abdominal segments divided into tergum and sternum. Segments 5–9 fused into complete rings

☐ Family: Leuctridae
see pages 29–30

3 The first segment much longer than the second and third segments: characteristic of a number of families (see below).

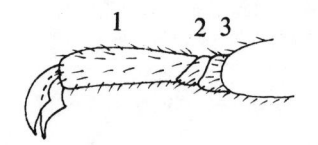

Examine carefully and determine:
(a) presence or absence of gills
(b) the size of the last segment of the maxillary palp (see diagram)

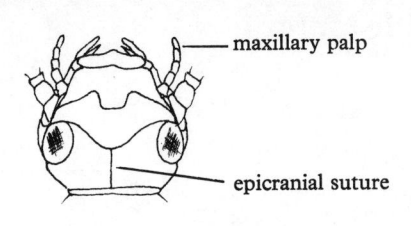

Key to families:

1 Gills in tufts on thorax near base of legs. Long hairs on posterior margins of leg segments

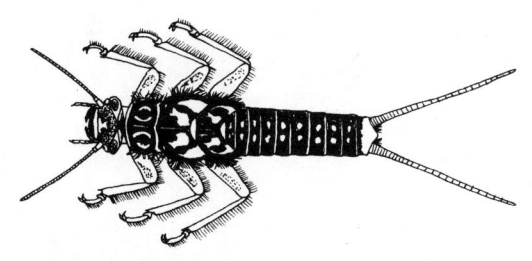

☐ Family: Perlidae
see page 30

Last segment of maxillary palp reduced and less than one-quarter as wide as preceding segment. Obvious long hairs on posterior margins of leg segments.

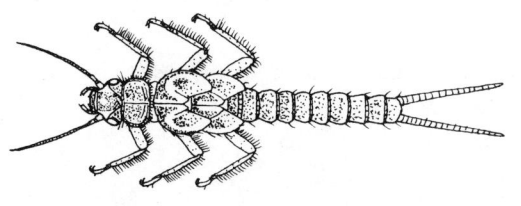

☐ Family: Chloroperlidae
see page 32

Gills absent
☐ 2

2 Last segment of maxillary palp normal and more than one-quarter as wide as preceding segment. Long hairs on posterior margins of leg segments. Gills absent

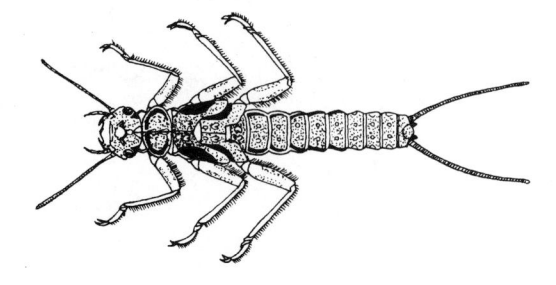

☐ Family: Perlodidae
see page 31

24

1 Family: Taeniopterygidae
(Examples)

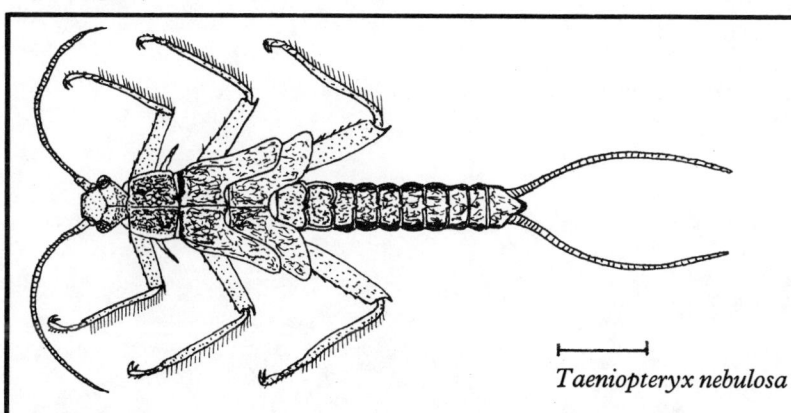

Taeniopteryx nebulosa

Description
First seven abdominal terga with horn-like processes on posterior edge.

Coxal region of legs (ventral view) bear a three-segmented filamentous gill.

Length: Up to 12 mm
Taeniopteryx nebulosa

Occurrence
Rivers and small stony streams. More characteristic of midland and northern areas

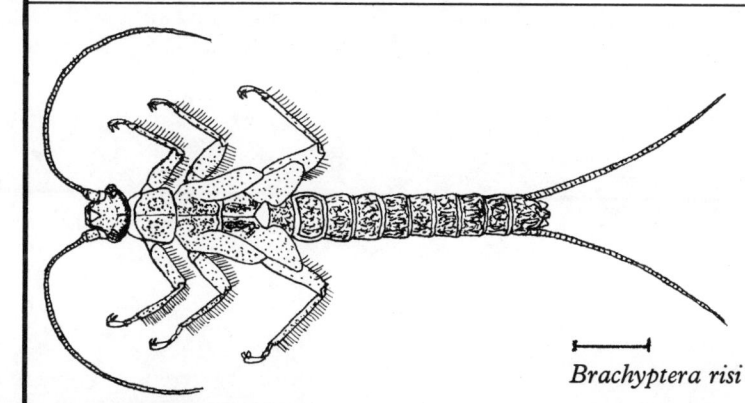

Brachyptera risi

Description
Short bristly hairs on abdominal terga. Basal segments of cerci bear long hairs on upper sides. No dark area at base of tibiae.
Length: Up to 10 mm
Brachyptera risi

Occurrence
Common in small stony streams

Family: Nemouridae

Examine carefully and determine:
(a) presence or absence of gills
(b) types of gills
(c) structure of segments of hind tarsus

Key to genera:

1 Gills present on prosternum
  ☐ 2
  Gills absent
  ☐ 3

2 Gills sausage-shaped and in 3 pairs

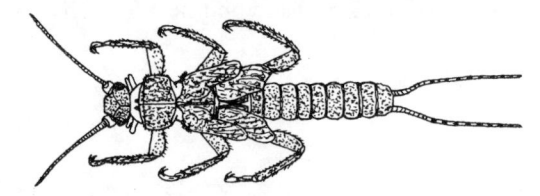

  ☐ Genus: *Protonemura*
       see page 27

  Gills filamentous, in two bunches of 5–8 gills on each side

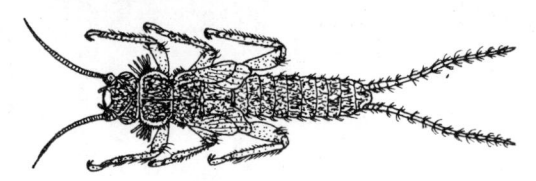

  ☐ Genus: *Amphinemura*
       see page 27

3 Segments 1 and 3 of hind tarsus about equal in length

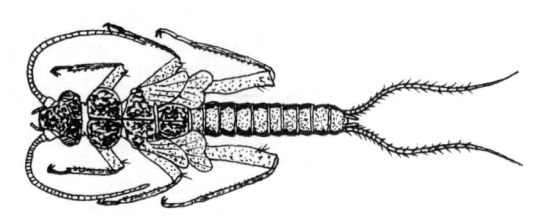

  ☐ Genus: *Nemurella*
       see page 28

Segment 1 of hind tarsus about half the length of segment 3

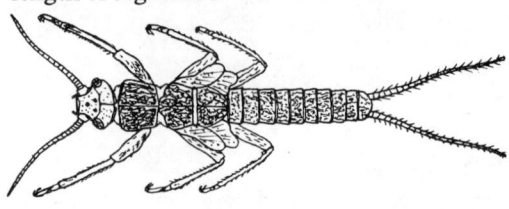

  ☐ Genus: *Nemoura*
       see page 28

Genus: *Protonemura*
(Example)

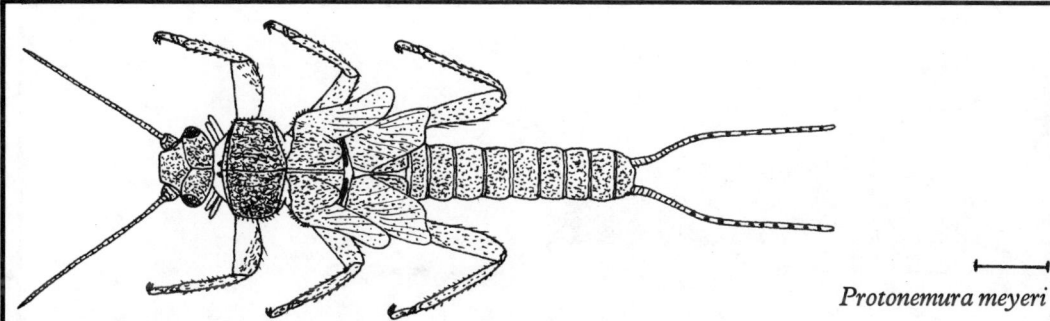

Protonemura meyeri

Description
First four abdominal segments divided into tergum and sternum. Only partial division of the segments 5–8.

Posterior margin of abdominal terga fringed with bristles. Dark patches on segments of cerci give them a ringed appearance

Colour: Dark green-grey
Length: Up to 10 mm
Protonemura meyeri

Occurrence
Swiftly flowing water, may be found at high altitudes

Genus: *Amphinemura*
(Example)

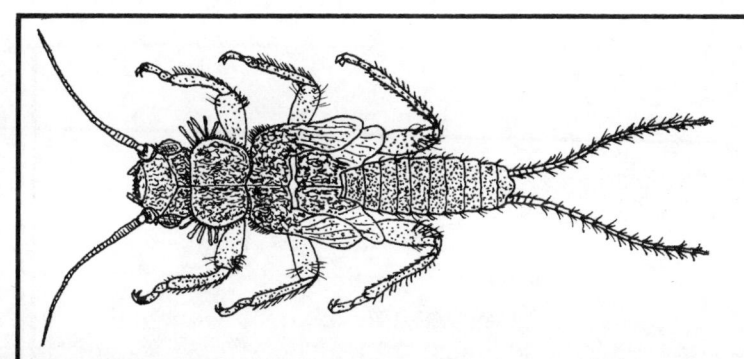

Amphinemura sulcicollis

Description
Long bristles present about two-thirds along length of each femur region of legs.

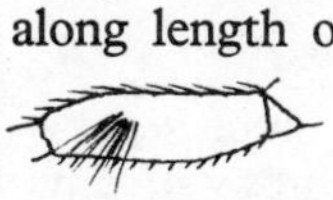

Terminal segments of cerci are elongated

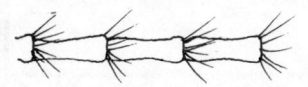

Colour: Dark brown, body often coated with debris
Length: Up to 6 mm
Amphinemura sulcicollis

Occurrence
Common in large streams and rivers with a stony substratum

Genus: *Nemurella*
(Example)

Nemurella picteti

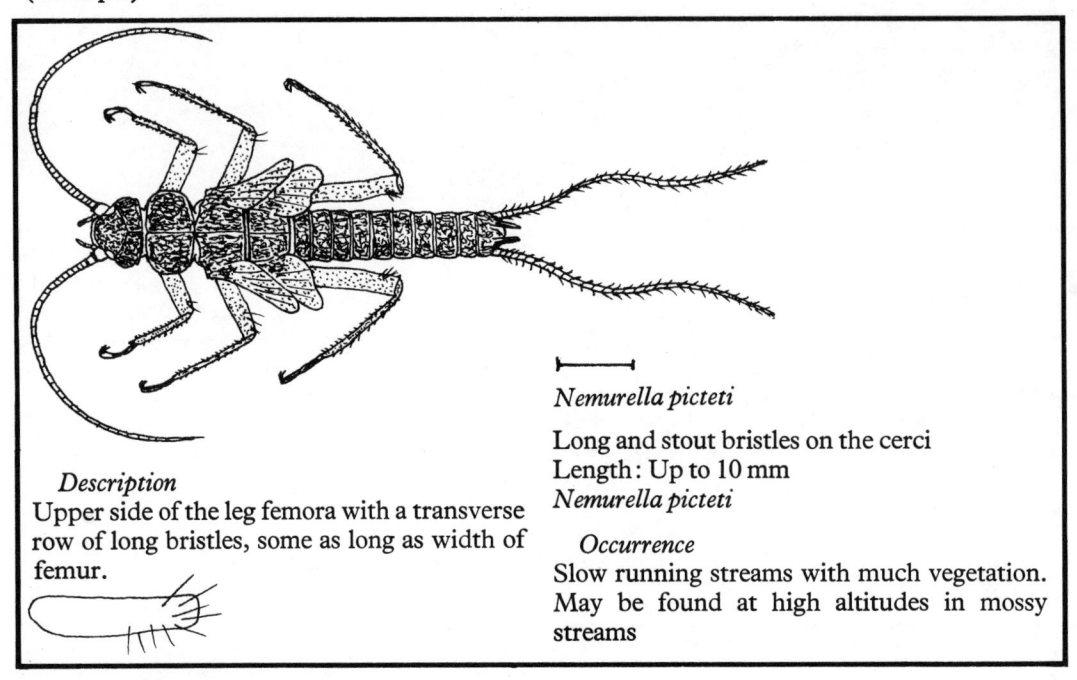

Long and stout bristles on the cerci
Length: Up to 10 mm
Nemurella picteti

Occurrence
Slow running streams with much vegetation. May be found at high altitudes in mossy streams

Description
Upper side of the leg femora with a transverse row of long bristles, some as long as width of femur.

Genus: *Nemoura*
(Example)

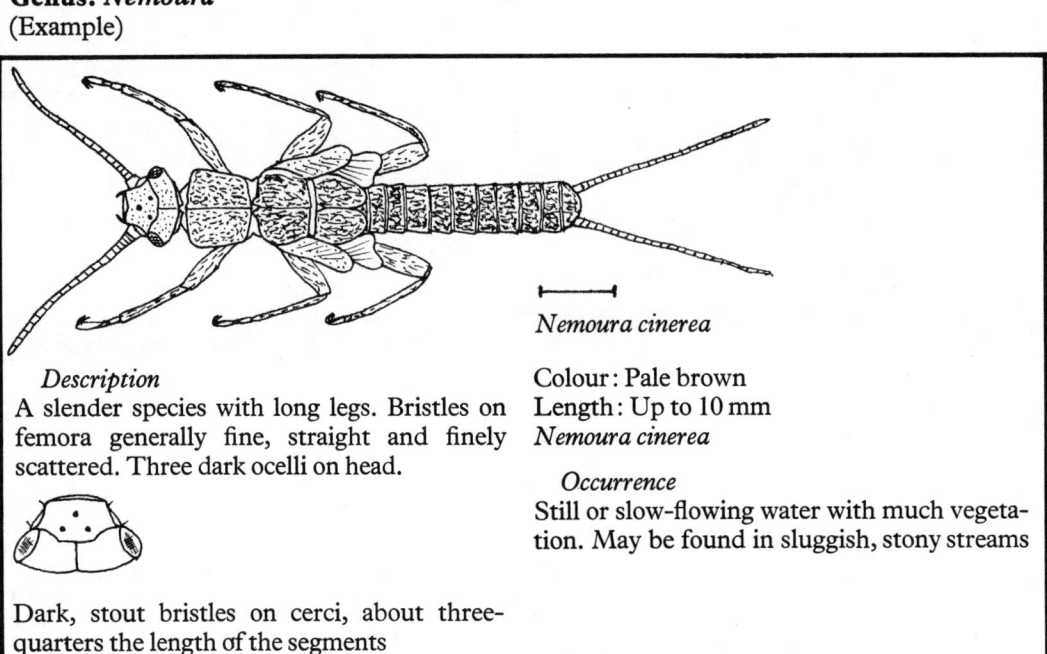

Nemoura cinerea

Colour: Pale brown
Length: Up to 10 mm
Nemoura cinerea

Occurrence
Still or slow-flowing water with much vegetation. May be found in sluggish, stony streams

Description
A slender species with long legs. Bristles on femora generally fine, straight and finely scattered. Three dark ocelli on head.

Dark, stout bristles on cerci, about three-quarters the length of the segments

Family: Leuctridae

Examine carefully and determine
(a) shape of head
(b) shape of antennae
(c) hairiness of body

Key to the more common species:

1 Head flattened and broad, antennae curved inwards
☐ *Leuctra geniculata*
Head rounded, antennae slender, not curved inwards
☐ 2

2 Body covered with long fine hairs
☐ *Leuctra nigra*
Body with few hairs
☐ *Leuctra hippopus*

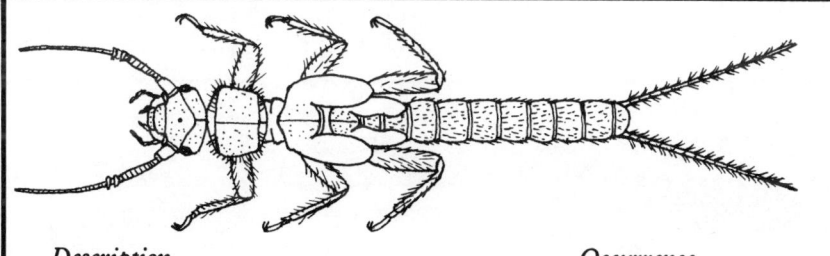

Leuctra geniculata

Description
A thick fringe of long hairs on the pronotum.
Colour: Brownish-grey
Length: Up to 12 mm
Leuctra geniculata

Occurrence
In streams and rivers with a stony substratum. More common in western and northern areas

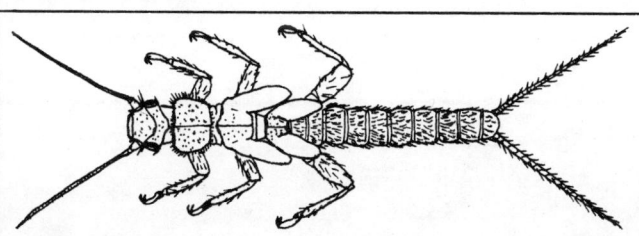

Leuctra nigra

Description
Colour: Yellow-orange
Length: Up to 8 mm
Leuctra nigra

Occurrence
Small stony streams with silty areas. More common in western and northern areas

Leuctra hippopus

Description
Colour: Greyish-yellow
Length: Up to 9 mm
Leuctra hippopus

Occurrence
Characteristic of stony streams and rivers

Family: Perlidae
(Examples)

(i) Pronotum more than twice as wide as long

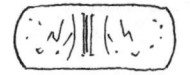

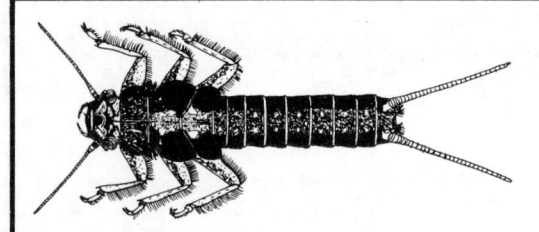

Dinocras cephalotes

Description
Colour: Dark brown or reddish, patterned with grey or yellow
Length: Up to 32 mm
Dinocras cephalotes

Occurrence
Characteristic of rivers with a stony substratum, sometimes in streams. Particularly common in western and northern areas

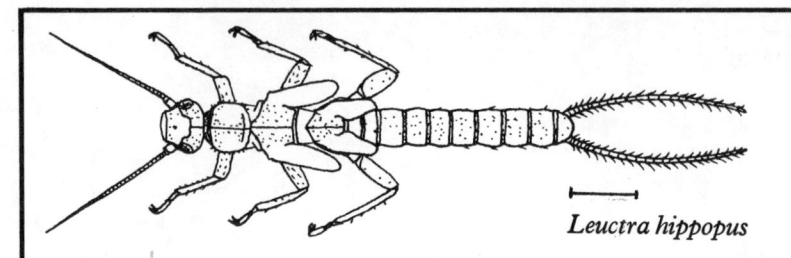

(ii) Pronotum less than twice as wide as long

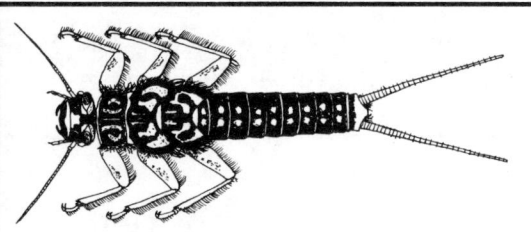

Perla bipunctata

Description
Colour: Black with yellow pattern
Length: Up to 34 mm
Perla bipunctata

Occurrence
Generally common in stony streams and rivers, particularly in western and northern areas

Family: Perlodidae
(Examples)

(i) First 4 abdominal segments divided into terga and sterna

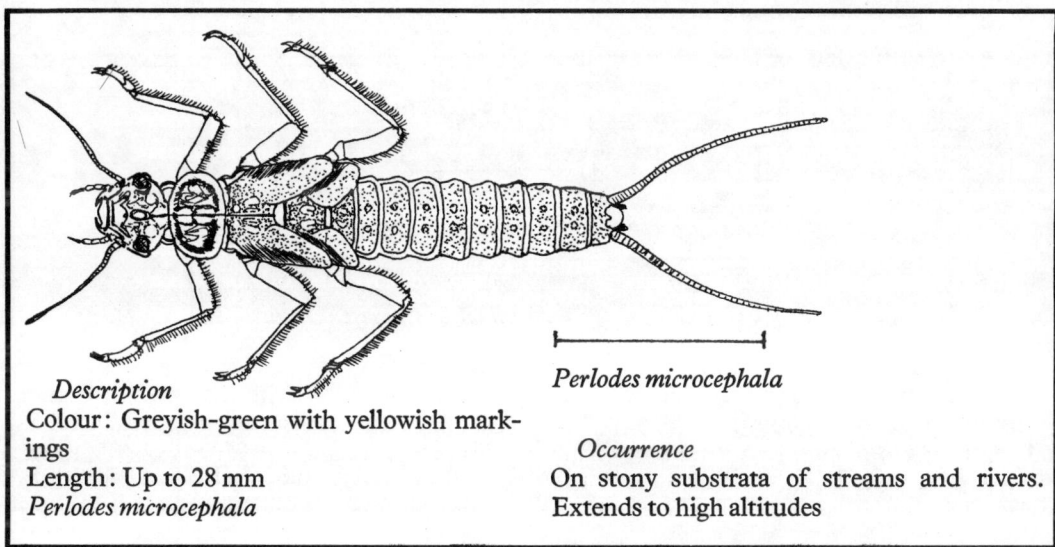

Perlodes microcephala

Description
Colour: Greyish-green with yellowish markings
Length: Up to 28 mm
Perlodes microcephala

Occurrence
On stony substrata of streams and rivers. Extends to high altitudes

(ii) First 2 abdominal segments divided into terga and sterna

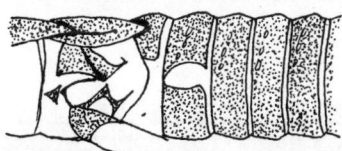

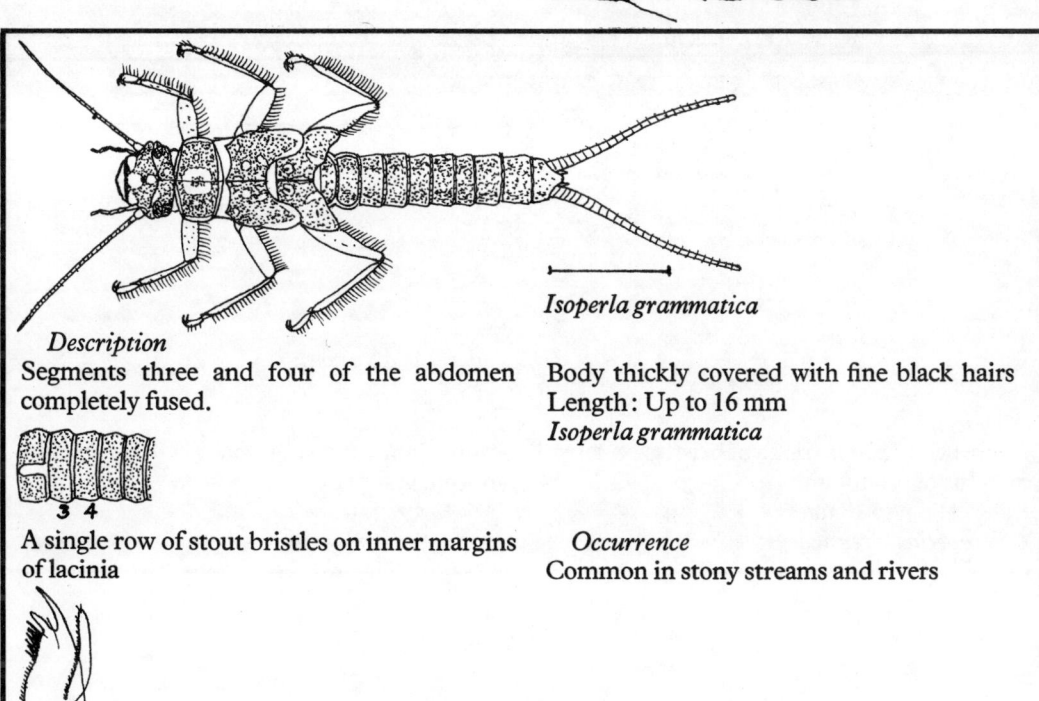

Isoperla grammatica

Description
Segments three and four of the abdomen completely fused.

A single row of stout bristles on inner margins of lacinia

Body thickly covered with fine black hairs
Length: Up to 16 mm
Isoperla grammatica

Occurrence
Common in stony streams and rivers

Family: Chloroperlidae
(Examples)

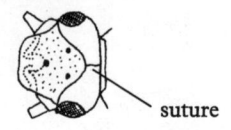

suture

(i) Epicranial suture obvious

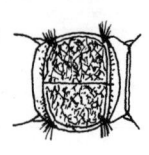

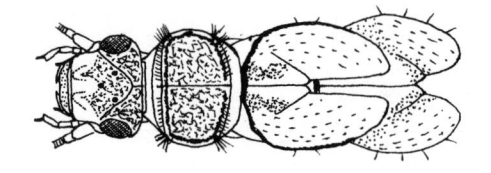

Chloroperla tripunctata

Description
Long fine bristles in groups on anterior and posterior areas of pronotum.
Length: Up to 10 mm
Chloroperla tripunctata

Occurrence
Streams and rivers with a stony substratum

(ii) Epicranial suture short or absent

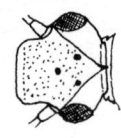

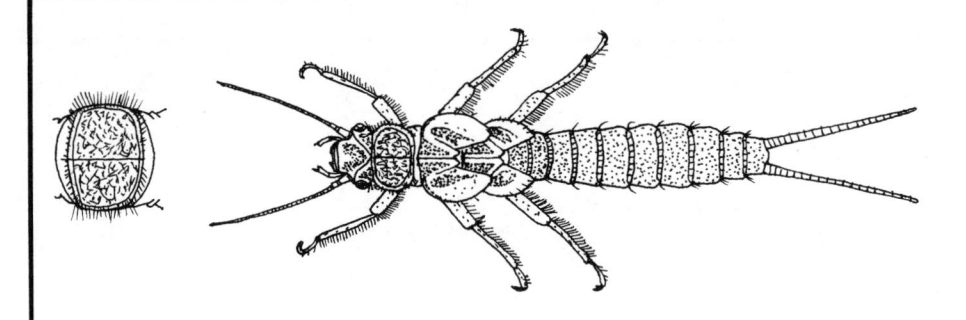

Chloroperla torrentium

Description
Almost continuous fringe of bristles on lateral margins of pronotum.
Length: Up to 10 mm
Chloroperla torrentium

Occurrence
Found in all types of water where the substratum is stony

Order: Ephemeroptera

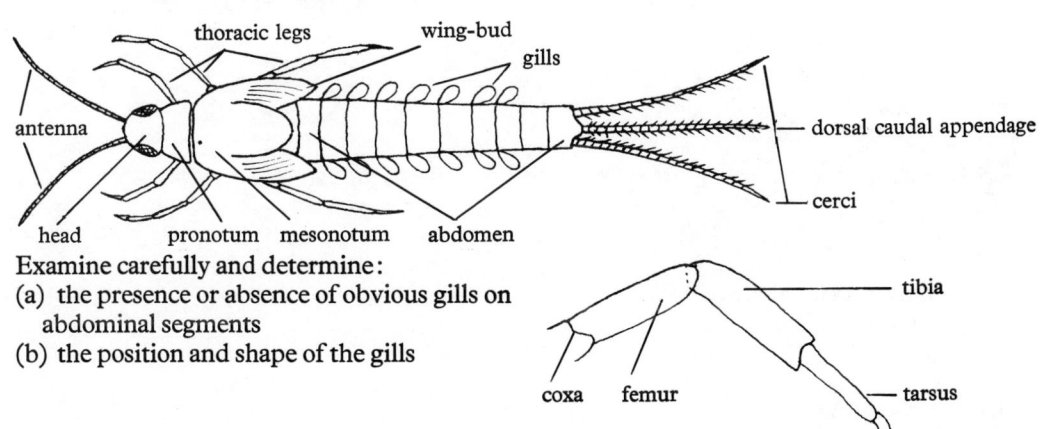

Examine carefully and determine:
(a) the presence or absence of obvious gills on abdominal segments
(b) the position and shape of the gills

Key to families:

1 Gills present, held on dorsal or lateral surfaces of abdominal segments
☐ 2

Gills not obvious. First pair reduced to thin filaments, second pair forming a large flap covering the first pair of gills

☐ Family: Caenidae
   see page 34

2 Gills held on dorsal surface of abdominal segments
☐ 3
Gills held on lateral surfaces
☐ 4

3 Gills very feathery, consists of two branches thickly fringed on both sides with fine filaments

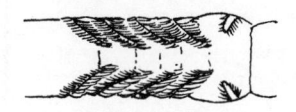

☐ Family: Ephemeridae
   see page 35

Gills not feathery, do not project beyond sides of abdominal segments

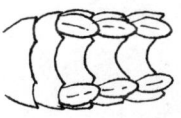

☐ Family: Ephemerellidae
   see pages 35–36

4 Gills plate-like in appearance, with a bunch of filaments. Body much flattened dorsoventrally

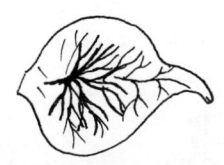

☐ Family: Ecdyonuridae
   see pages 36–38

Gills may be plate-like but without a bunch of filaments
☐ 5

5 Gills typically with obvious projections from the free end. Body long and round

☐ Family: Leptophlebiidae
see pages 39–40

Gill plates shaped like a heart, a tennis racket head or a beech leaf

☐ Family: Baetidae
see pages 41–43

Family: Caenidae
(Examples)

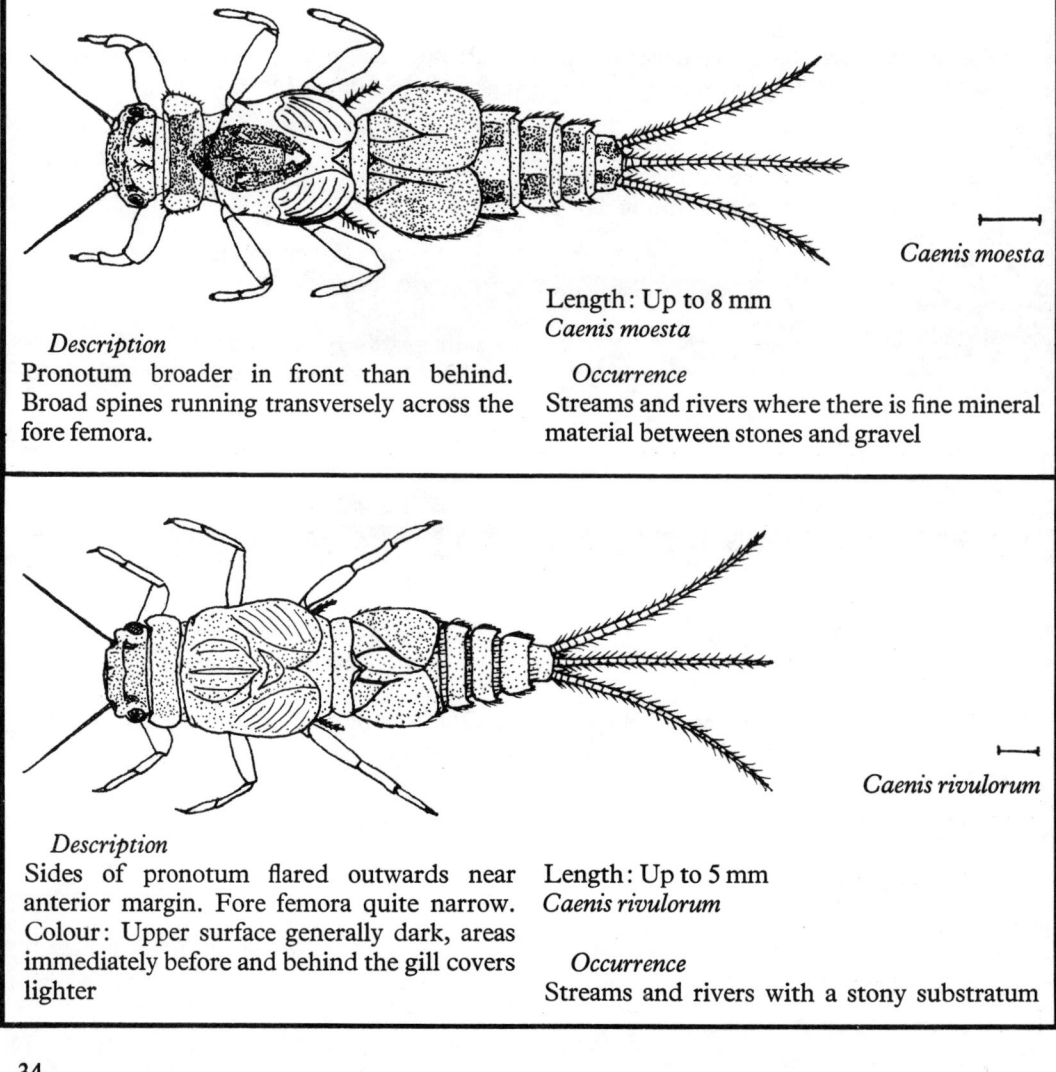

Caenis moesta

Length: Up to 8 mm
Caenis moesta

Description
Pronotum broader in front than behind. Broad spines running transversely across the fore femora.

Occurrence
Streams and rivers where there is fine mineral material between stones and gravel

Caenis rivulorum

Description
Sides of pronotum flared outwards near anterior margin. Fore femora quite narrow. Colour: Upper surface generally dark, areas immediately before and behind the gill covers lighter

Length: Up to 5 mm
Caenis rivulorum

Occurrence
Streams and rivers with a stony substratum

Family: **Ephemeridae**
(Examples)

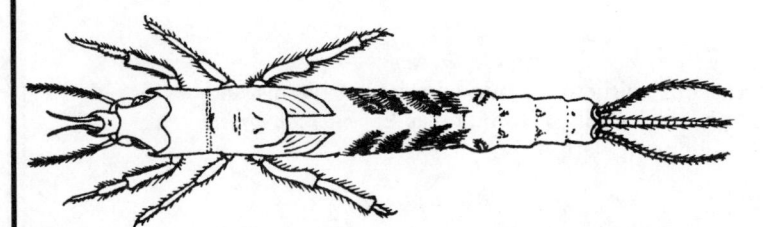

Ephemera danica

Description
Mandibles extend beyond the head. Large dark triangular markings on abdominal terga 7–9.

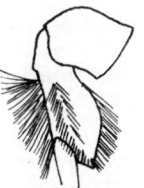

Tibia of fore-legs broad
Length: Up to 25 mm
Ephemera danica

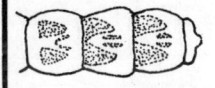

Occurrence
In lakes, rivers and streams particularly if the substratum is gravelly or sandy

Ephemera vulgata

Description
Similar in appearance to *Ephemera danica*, but with distinct markings on both sides of most abdominal terga.

Tibia of fore-legs narrow
Length: Up to 25 mm
Ephemera vulgata

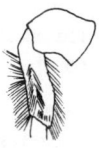

Occurrence
Rivers and streams with a muddy substratum. Not as common as *Ephemera danica*

Family: **Ephemerellidae**
(Examples)

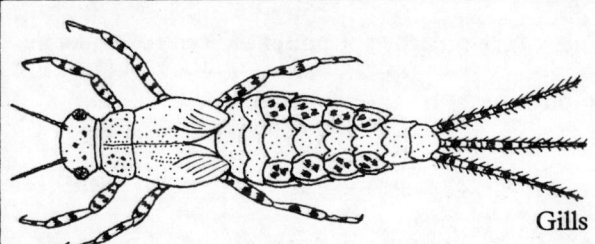

Ephemerella ignita

Description
Tails with alternating light and dark bands. Tubercles (humps) present on the posterior margins of the abdominal terga.

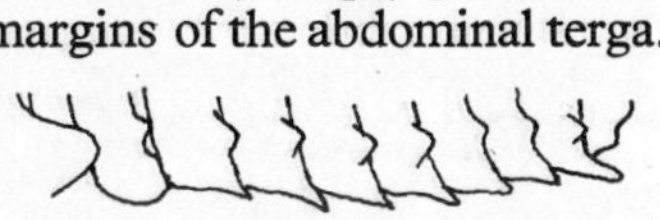

Gills marked with a clover leaf like pattern
Length: Up to 10 mm 
Ephemerella ignita

Occurrence
May be found in streams and rivers where vegetation is thick. Also found in streams and rivers under stones where the current is fast

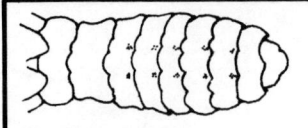

Ephemerella notata

Description
Similar to *Ephemerella ignita* but without alternating light and dark bands on the tails. Tubercles not conspicuous. Gills with dark area similar in shape to the gill.
Length: Up to 10 mm
Ephemerella notata

Occurrence
Found in stony streams and rivers. Not as common as *Ephemerella ignita*

Family: Ecdyonuridae

Nymphs of this family have flattened bodies and legs, and gills comprising a flattened lamella or plate and typically a tuft of filaments. The nymphs are usually found clinging to stones and boulders within the water. Examine carefully and determine

(a) the shape of the hind corners of the pronotum
(b) presence or absence of a dark spot in the centre of the femur of each leg

1 Hind margin of pronotum directed backwards on each side of mesonotum. No dark spot at centre of femora

☐ Genus: *Ecdyonurus*
(**Note**: projections not present in very young nymphs).*
see page 37

No backwardly directed projections on pronotum
☐ 2

2 Distinct dark spot present at centre of femora

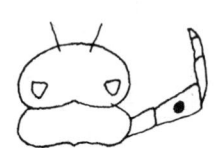

☐ Genus: *Rhithrogena*
see page 37

No dark spot present at centre of femora

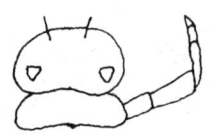

☐ Genus: *Heptogenia*
see page 38

* Young nymphs of the genus *Ecdyonurus* may be confused with members of the genus *Heptogenia*. However, although members of both genera have flattened bodies and legs, members of the genus *Ecdynorus* are extremely broad and flattened.

Genus: *Ecdyonurus*
(Example)

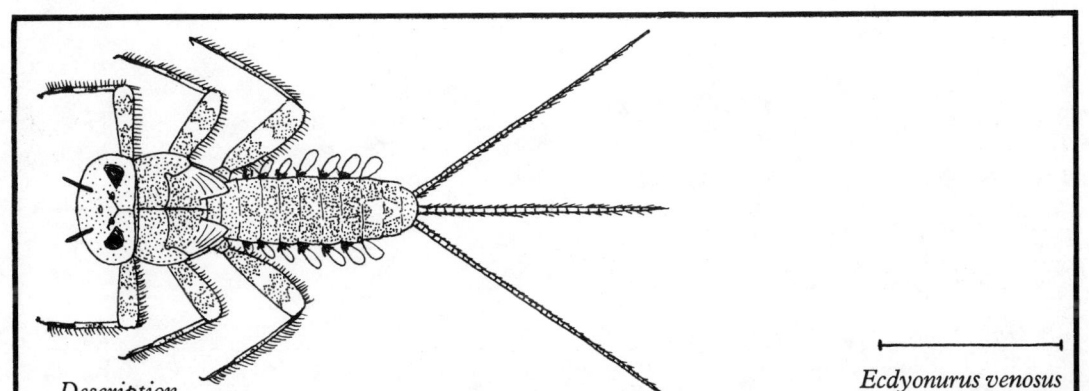

Ecdyonurus venosus

Description
Last gill without filaments. Sides of pronotum slightly curved. Tarsi dark at apex. Claws of legs with less than three teeth.
Length: Up to 25 mm
Ecdyonurus venosus

Occurrence
Fairly common in streams and rivers with a stony substratum and fast flow. *Ecdyonurus torrestis* is usually found in smaller streams.

Similar to *Ecdyonurus venosus* but tarsi dark at apex and base, and three or more teeth on the claws of the legs.
Length: Up to 25 mm
Ecdyonurus torrentis

Genus: *Rhithrogena*
(Example)

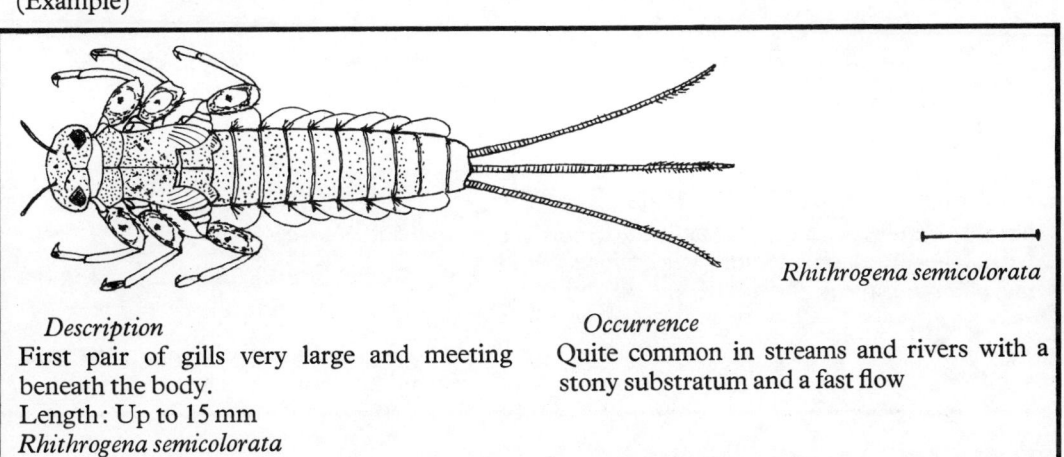

Rhithrogena semicolorata

Description
First pair of gills very large and meeting beneath the body.
Length: Up to 15 mm
Rhithrogena semicolorata

Occurrence
Quite common in streams and rivers with a stony substratum and a fast flow

Genus: *Heptogenia*

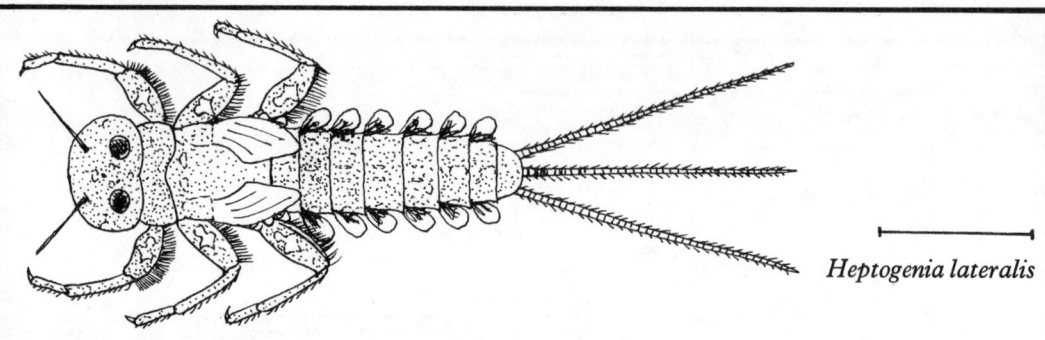

Heptogenia lateralis

Description
First gill small. Lamella of gill larger than filaments. 7th pair of gills without filaments. Femora with a light cross-shaped area. Front edge of femora fringed with stout hairs.

Length: Up to 20 mm
Heptogenia lateralis

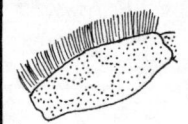

Occurrence
Found in streams with a stony substratum

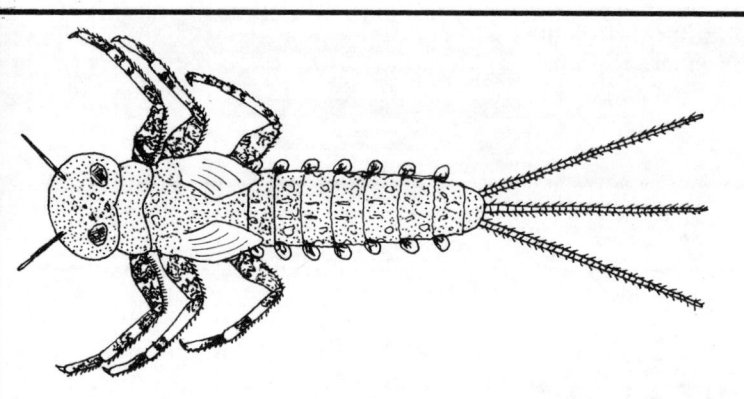

Heptogenia sulphurea

Description
Gills small and rounded at the tip of the lamella. Femora with distinct dark transverse bands. Front edge of femora with fringe of fine hairs and spines.

Colour: Patterned black and white
Length: Up to 20 mm
Heptogenia sulphurea

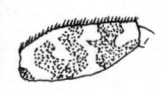

Occurrence
Common in large stony streams and rivers. Can extend into slower reaches

Family: Leptophlebiidae

Examine the gills closely.

Key to genera:

1 Gills divided into several branches

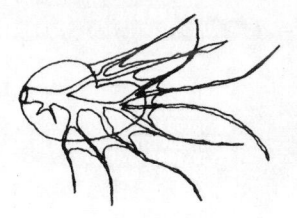

☐ Genus: *Habrophlebia*
see page 39

Gills divided into two branches
☐ 2

2 Branches of gills straight-sided

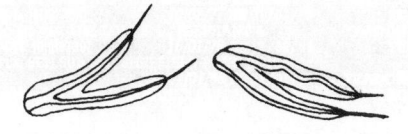

☐ Genus: *Paraleptophlebia*
see page 40

All except first pair of gills with branches expanded at the base

☐ Genus: *Leptophlebia*
see page 40

Genus: *Habrophlebia*
(Example)

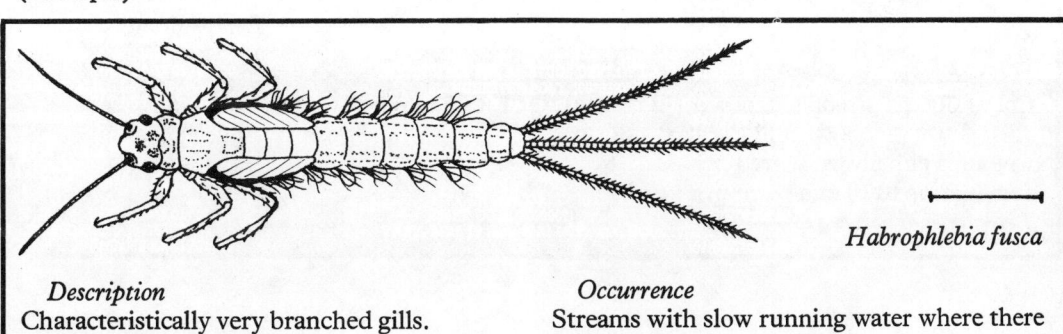

Habrophlebia fusca

Description
Characteristically very branched gills.
Colour: Dark reddish-brown
Length: Up to 15 mm
Habrophlebia fusca

Occurrence
Streams with slow running water where there is vegetation and plant debris

Genus: *Paraleptophlebia*
(Example)

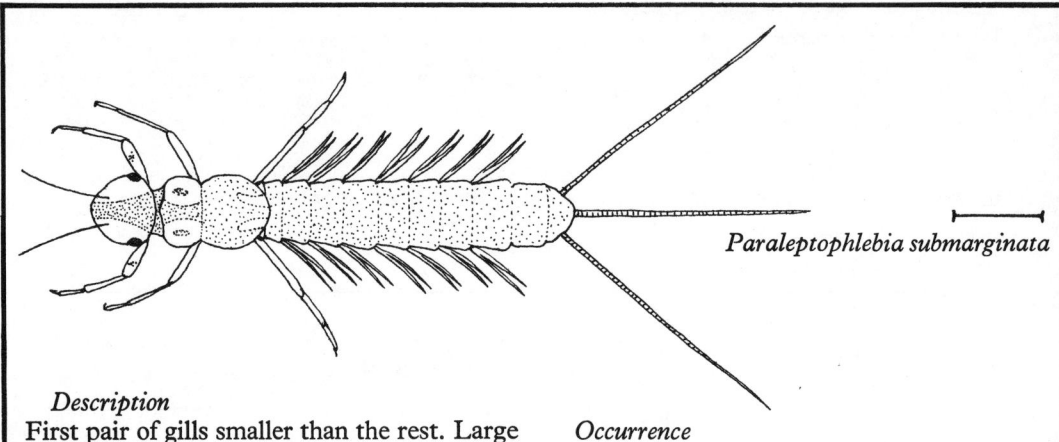

Paraleptophlebia submarginata

Description
First pair of gills smaller than the rest. Large maxillary palps, last segment densely covered with long hairs.
Colour: Dark brownish-red
Length: Up to 12 mm
Paraleptophlebia submarginata

Occurrence
In streams and rivers with slow flowing water

Genus: *Leptophlebia*
(Example)

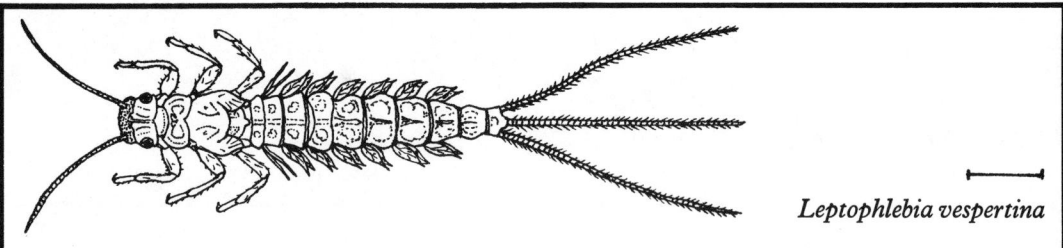

Leptophlebia vespertina

Description
Claws of legs strongly toothed. Lamellae of gills gently merge with filament at the tip.
Colour: Dark brownish-red
Length: Up to 10 mm
Leptophlebia vespertina

Occurrence
Slow flowing streams, ponds and lakes

Family: Baetidae

Examine carefully and determine
(a) length of the 'tails' and their markings
(b) the shape of the gills

Key to genera:

1 All three tails approximately the same length, marked with dark rings and often a dark band

☐ 2

Middle tails shorter than the outer ones, no dark rings, black band may be present. Gills single and rounded at tip

☐ Genus: *Baetis*
   see pages 41–42

2 First six pairs of gills double

☐ Genus: *Cloeon*
   see page 42

Gills single
☐ 3

3 Gills shaped like a beech leaf. About five dark rings on tails between dark band and body

☐ Genus: *Centroptilum*
   see page 42

Gills not shaped like a beech leaf. More than five dark rings on tails between dark band and body

☐ Genus: *Procloeon*
   see page 43

Genus: *Baetis*
(Examples)

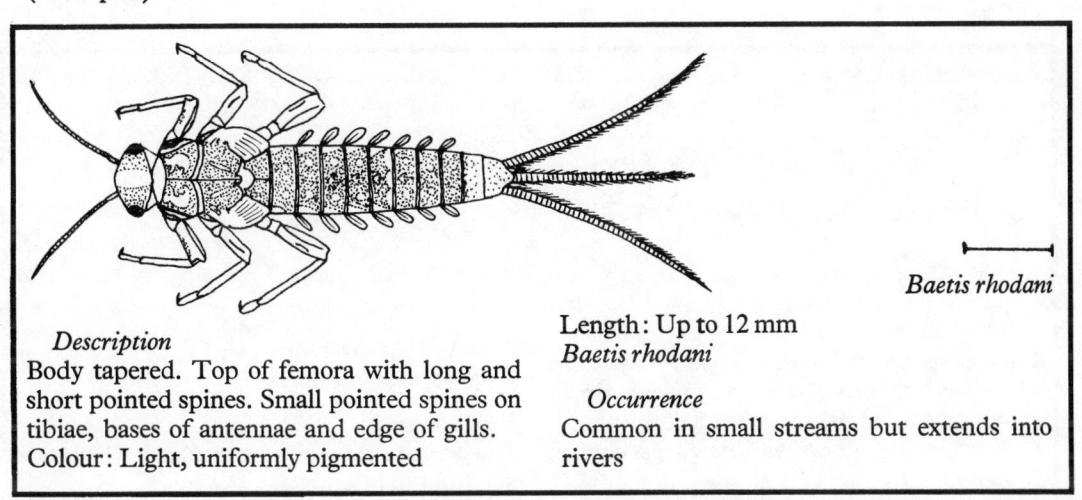

Baetis rhodani

Description
Body tapered. Top of femora with long and short pointed spines. Small pointed spines on tibiae, bases of antennae and edge of gills.
Colour: Light, uniformly pigmented

Length: Up to 12 mm
Baetis rhodani

Occurrence
Common in small streams but extends into rivers

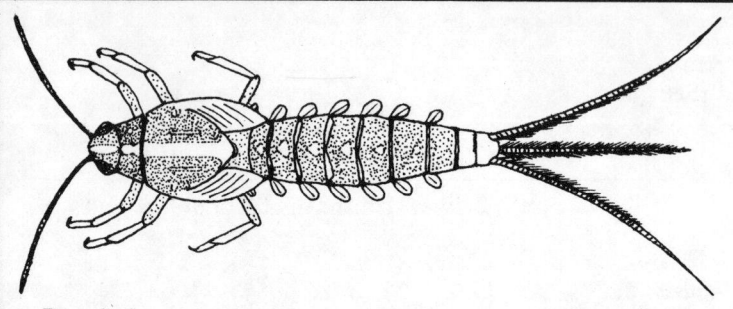

Baetis muticus

Description
Body only slightly tapered. Long pointed spines on femora and tibiae. Gills rather elongated with a toothed margin.
Colour: Dark
Length: Up to 12 mm
Baetis muticus

Occurrence
Typical of small streams with a stony substratum but extends into rivers

Genus: *Cloeon*
(Example)

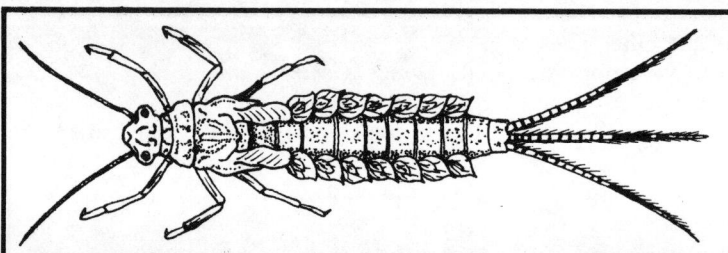

Cloeon simile

Description
Dark rings and a dark band on the tails. Bands broad. Hairs fringing tails not very obvious. Gills pointed, second lamella small.
Colour: Light coloured, strongly contrasting pattern

Length: Up to 11 mm
Cloeon simile

Occurrence
Found in the slower parts of streams and rivers, and also in ponds and weedy edges of lakes

Genus: *Centroptilum*
(Example)

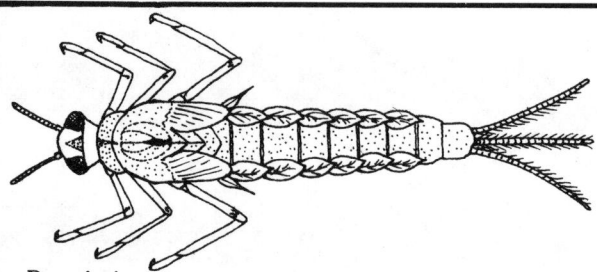

Centroptilum luteolum

Description
Dark rings but no dark band on tails. Gills single and shaped like a beech leaf. Small spines on abdominal segments 8 and 9.
Colour: Sandy with a variegated pattern on the abdomen

Length: Up to 8 mm
Centroptilum luteolum

Occurrence
In streams and rivers particularly where the substratum is sandy

Genus: *Procloeon*
(Example)

Procloeon pseudorufulum

Description
Dark rings and a dark band on the tails. Tails fringed with thick hairs. Dark band narrow and beyond the middle of each tail. Colour: Sandy with no strong contrast between light and dark areas

Length: Up to 9 mm
Procloeon pseudorufulum

Occurrence
In streams and rivers particularly where the substratum is sandy

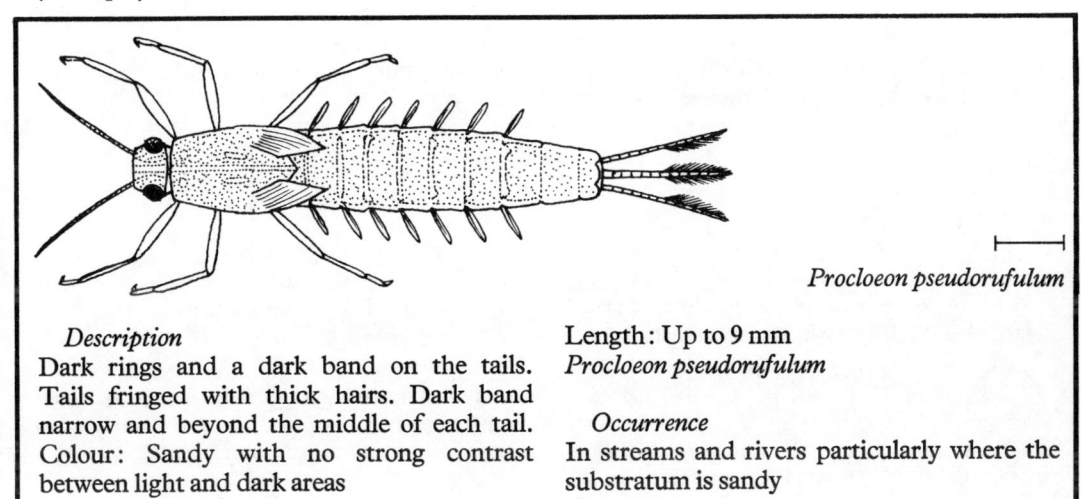

Order: Odonata (damselflies and dragonflies)

Key to sub-orders:

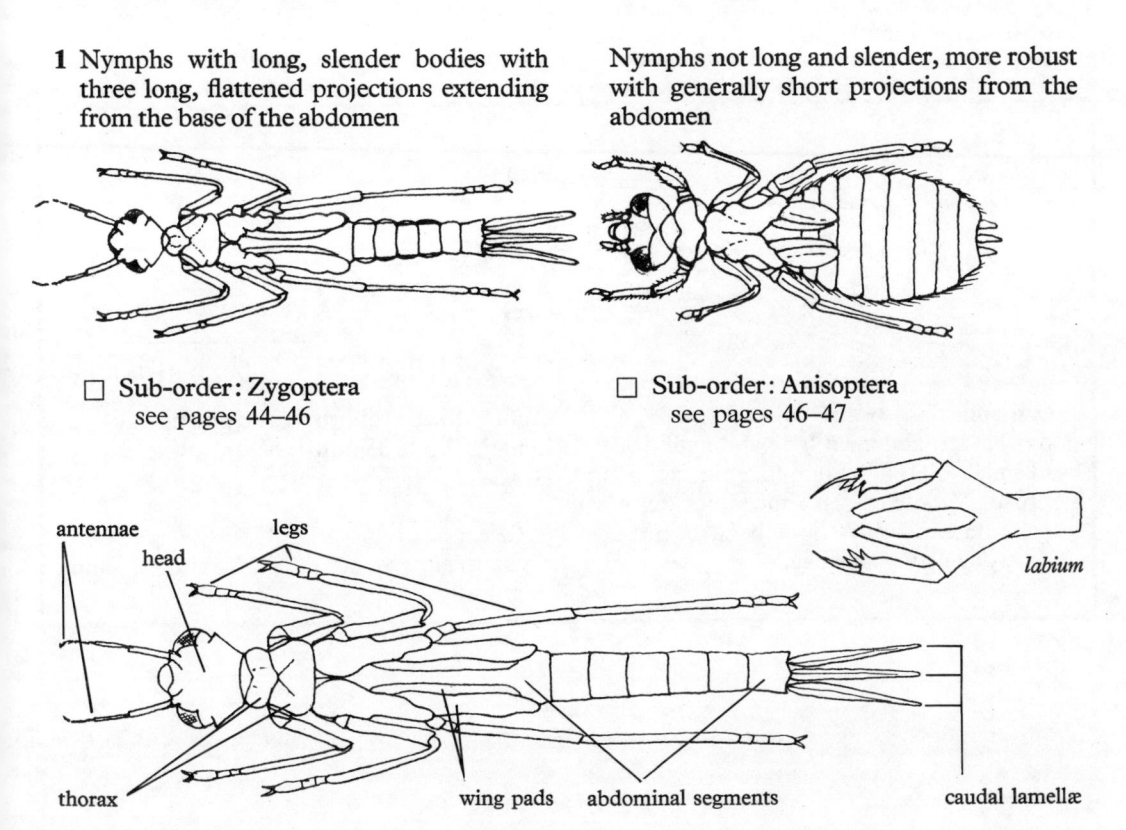

1 Nymphs with long, slender bodies with three long, flattened projections extending from the base of the abdomen

Nymphs not long and slender, more robust with generally short projections from the abdomen

☐ Sub-order: Zygoptera
   see pages 44–46

☐ Sub-order: Anisoptera
   see pages 46–47

labium

antennae
head
legs
thorax
wing pads
abdominal segments
caudal lamellæ

43

Sub-order: Zygoptera

Examine carefully and determine
(a) shape of the labium (mask)
(b) the shape of the caudal lamellae

Key to families:

1 Labium wide with a deep cleft. Caudal lamellae thick and triangular in section

☐ Family: Agriidae
  see pages 44–45

Labium without a deep central cleft
☐ 2

2 Labium with a projecting central lobe. Caudal lamellae with long narrow pointed tips. Nymphs rather short and broad

☐ Family: Platycnemididae
  see pages 45–46

Labium triangular. Caudal lamellae short and broad

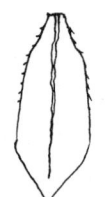

☐ Family: Coenagriidae
  see page 46

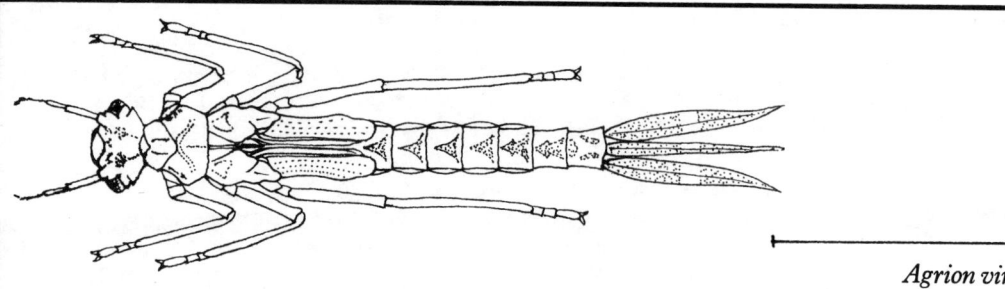

Family: Agriidae
(Examples)

Agrion virgo

Description
Body slender, abdomen long and cylindrical, head small. Cleft of labium four times as long as broad. Middle caudal lamella not quite as long as outer ones. A light band is present mid-way along the lamellae.

Colour: Shades of brown and green
Length: Up to 35 mm
Agrion virgo

Occurrence
Swift streams where the substratum is sandy or gravelly

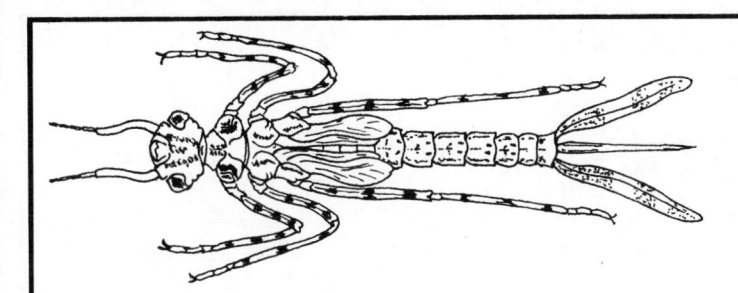

Agrion splendens

Description
Body slender, head small and round. Cleft of labium less than four times as long as broad. Middle caudal lamella distinctly shorter than laterals. Lamellae usually with two light bands mid-way.

Colour: Variable—red-brown, grey, bright or dull brown
Length: Up to 45 mm
Agrion splendens

Occurrence
Quiet streams on mud or amongst vegetation

Family: Platycnemididae
(Examples)

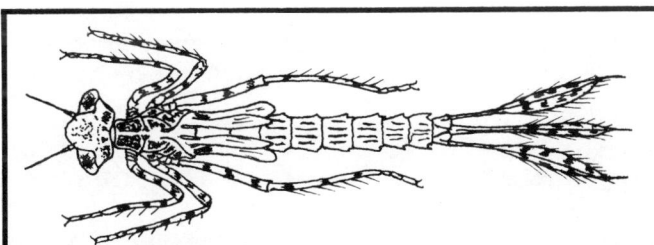

Platycnemis pennipes

Description
Head large, 5-sided, eyes large. Labium long and rather triangular. Lateral lobes with one long and one short hook. Sides of abdomen with spines. Caudal lamellae long and widest near the tip.

Colour: Brown-yellow, spotted, spiny
Length: Up to 22 mm
Platycnemis pennipes

Occurrence
In running water amongst vegetation

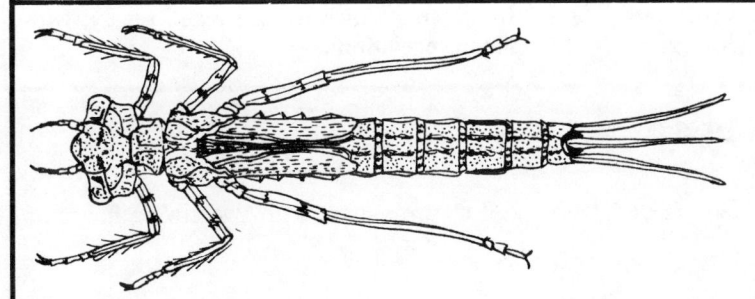

Pyrrhsoma nymphula

Description
Body appears stumpy. Head wide with large eyes. Labium triangular with one long and one short hook on lateral lobes. Abdomen tapers to a square end. Caudal lamellae paddle-shaped and pointed at tip.

Colour: Usually dark brown, lamellae spotted brown, legs brown banded
Length: Up to 23 mm
Pyrrhsoma nymphula

Occurrence
In streams particularly on leaves and debris

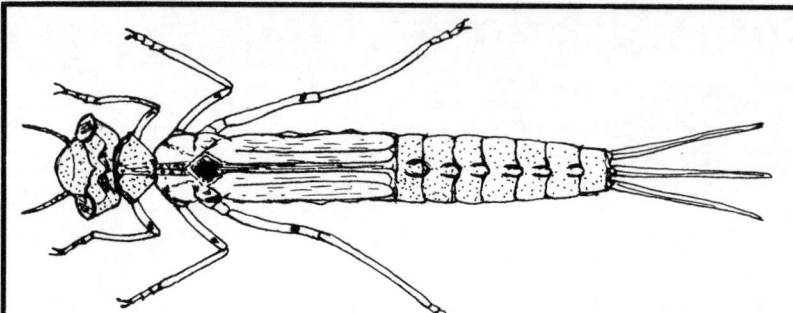

Enallagma cyathigerum

Description
Abdomen long and rather stout, head wide and short, 5-sided. Labium triangular with one long and one short curved hook on laterals. Caudal lamellae leaf-like and rather short with dark line across the middle.

Colour: Bright green, sometimes brown. A pale spot surrounded by a dark ring on each abdominal segment
Length: Up to 28 mm
Enallagma cyathigerum

Occurrence
In quiet streams amongst vegetation

Family: Coenagriidae
(Example)

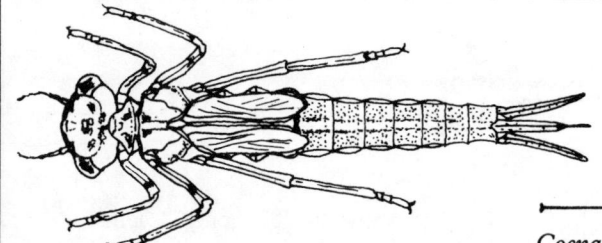

Coenagrion mercuriale

Description
Body slender, abdomen tapering. Labium triangular with one long and one short hook on laterals. Caudal lamellae short, broad and shortly pointed.
Colour: Greenish-yellow with dark spots down abdomen. Legs pale in colour

Length: Up to 18 mm
Coenagrion mercuriale

Occurrence
In streams which are clear and boggy, with much vegetation

Sub-order: Anisoptera

1 Labium triangular and spoon-shaped, with triangular laterals

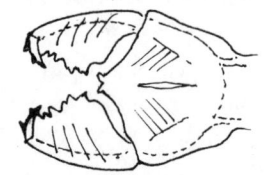

☐ Family: Cordulegasteridae
see page 47

2 Labium square with two strong hooks on the laterals

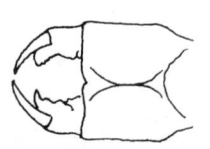

☐ Family: Gomphidae
see page 47

Family: Cordulegasteridae
(Example)

Cordulegaster boltonii

Colour: Brown
Length: Up to 42 mm
Cordulegaster boltonii

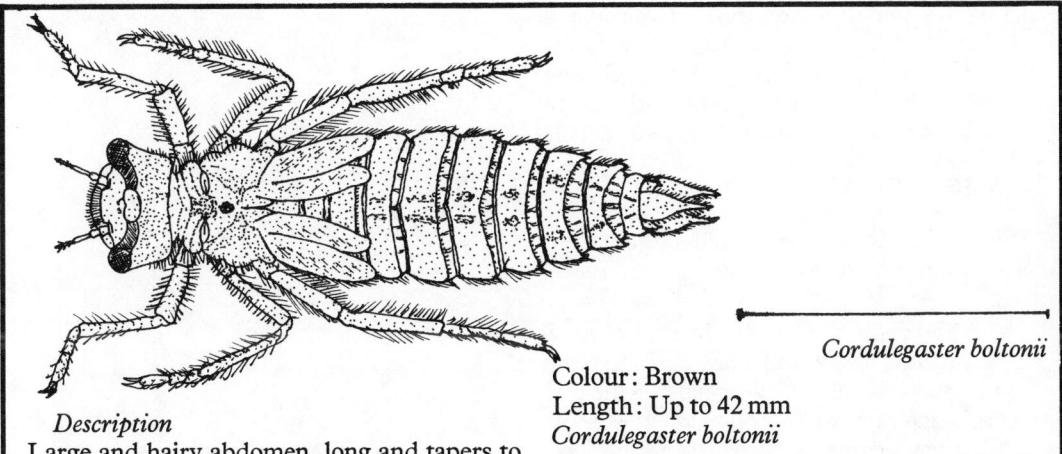

Description
Large and hairy abdomen, long and tapers to the tip. Head broad and rather flat, eyes small. Labium triangular and spoon-shaped. Lateral lobes with toothed edges.

Occurrence
Characteristic of pools in rapid streams and rivers, living under debris

Family: Gomphidae
(Example)

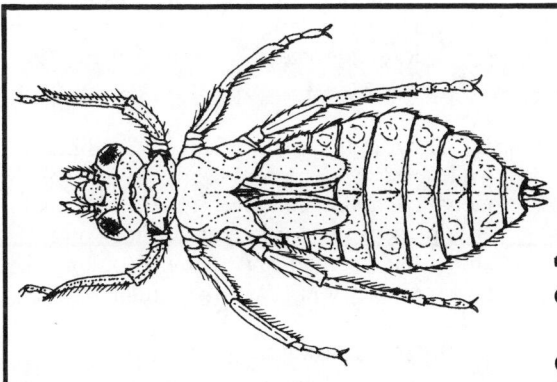

Gomphus vulgatissimus

Colour: Brown
Length: Up to 30 mm
Gomphus vulgatissimus

Description
Flattened, oval in shape, hairy. Head small and heart-shaped, antennae club-like. Labium square with two strong hooks on laterals. Caudal lamellae very short.

Occurrence
Streams and rivers where there is sand and mud

Class: Insecta—Larvae

Some confusion could arise when attempting to identify some of the larvae of the insect orders described here. Care needs to be taken when attempting to identify the specimens collected. If any confusion should arise refer to the later sections which give greater detail.

Key to orders:

1 Three pairs of well-developed and obvious legs on the thoracic segments
☐ 2

Without distinct thoracic legs. Legs, if present, are fleshy and may be found on both thoracic and abdominal segments. Head may be well-developed. Posterior abdominal segments with various projections
☐ Order: Diptera
  see pages 49–51

2 Head well-developed, antennae may be quite long. On some species gills are present on the abdominal segments. Tail processes of various designs extend from the last abdominal segment
☐ 3

Head and thoracic segments usually heavily sclerotized (hardened). Many possess gills either on the abdominal segments, as anal gills at the end of the abdomen, or both. May live in a case made of a variety of different materials
☐ Order: Trichoptera
  see pages 52–63

3 Specimens usually less than 15 mm long, some are very small. Head well-developed but without obvious mandibles. Some with long, leg-like or spiny processes extending from the abdominal segments. Tail processes of various designs
☐ Order: Coleoptera
  see page 64

4 Usually up to 20 mm long. Mouthparts well-developed, mandibles large and obvious. A pair of long, jointed, leg-like gills present on the lateral surfaces of the abdominal segments. Abdomen prolonged into a single, long tapering caudal appendage
☐ Order: Neuroptera (alder flies)

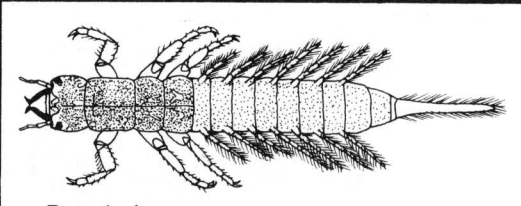

Order: Neuroptera
Example: *Sialis*

Sialis lutaria

Description
Colour: Brownish
Length: Up to 22 mm long
Sialis lutaria
Note: May be confused with the larva of Gyrinus, see section on Coleoptera larvae

Occurrence
Common in ponds and very slow-flowing water of streams and rivers where the substratum consists of mud and vegetation is present

Order: Diptera—Larvae

The insect order Diptera contains many species, and a large number of the larval stages are aquatic. The larvae may be active swimmers, live in the mud and sand deposits or among the vegetation. Some species live in specially constructed mud or sand tubes.

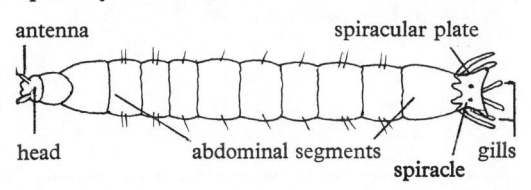

Accurate identification of the many Diptera larvae likely to be encountered in a thorough survey of streams and rivers is beyond the scope of this book. In the main only those families and some common examples of larvae likely to be encountered in any survey are considered here.

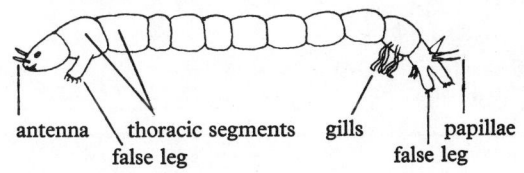

Examine carefully and determine
(a) shape of the body
(b) development of the head
(c) presence or absence of obvious appendages

Key to families:

1 Body fat and maggot-like, rather soft and fleshy
 ☐ 2
 Body maggot or worm-like but not fat and fleshy
 ☐ 5

2 Head usually well-developed and obvious
 ☐ 3
 Head not well-developed, not obvious or very small
 ☐ 4

3 Fleshy false legs may be present. Last abdominal segment with a large spiracular plate and 3 or 4 pairs of lobes (gills)

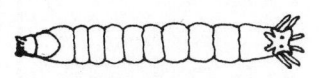

 ☐ Family: Tipulidae
  see page 50

Body dumb-bell shaped, swollen at the posterior end. Larvae typically attached to stones where current is swift. Head equipped with a fringe of fine hair-like structures

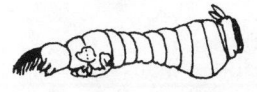

 ☐ Family: Simuliidae
  see page 50

4 Head minute, retractile. Short fleshy protuberences present as extensions of the abdominal segments

 ☐ Family: Tabanidae
  see page 51

Head not visible, anterior segments much tapered. A pair of black mouth hooks visible at the anterior end

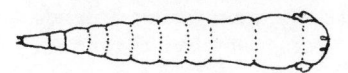

 ☐ Family: Anthomyiidae
  see page 51

5 Head dark and well-developed. A pair of prolegs present on the first thoracic segment. Sausage-shaped anal gills present on last and sometimes the last but one, abdominal segment

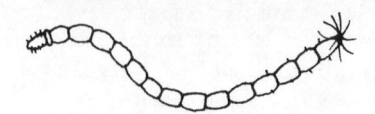

☐ Family: Chironomidae
see page 51

Body long and thin, worm- or snake-like. Prolegs absent. Usually with a fan of hair-like structures on the last abdominal segment

☐ Family: Ceratopogonidae

Family: Tipulidae
(Examples)

Dicranota sp.

Description
A pair of cylindrical false legs on the abdominal segments 3–7, each ends in a circlet of hooks.
Colour: Whitish
Length: Up to 20 mm
Genus: *Dicranota*

Occurrence
In the mud of streams and rivers

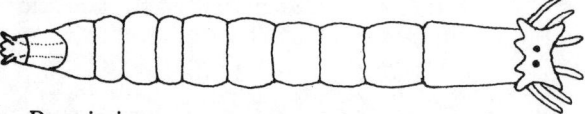

Tipula sp.

Description
Head retractile. Last abdominal segment with a rectangular lobed spiracular plate. Six sausage-shaped gills near the plate.

Length: Up to 30 mm—may be longer if fully extended
Genus: *Tipula*

Occurrence
Common under stones or in the mud of shallow streams

Family: Simuliidae
(Example)

Simulium sp.

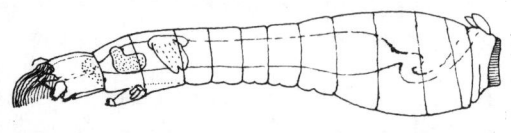

Head with a pair of prominent mouth brushes. Hooked false leg on thorax. Abdomen ends in an attachment organ of radial rows of strong hooks.

Colour: Creamy white or grey
Length: Largest species up to 20 mm, others about 10 mm
Genus: *Simulium*

Occurrence
On stones, debris and vegetation in all types of water courses from trickles to large rivers

Family: Tabanidae
(Example)

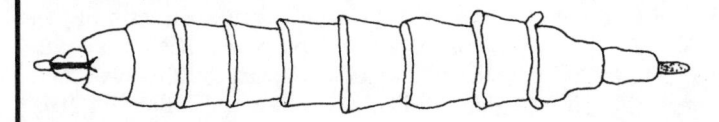

Tabanus sp.

Description
Last abdominal segment tapering into extensible siphon.
Colour: White or pale yellow

Length: Up to 40 mm
Genus: *Tabanus*
Occurrence
Common in streams among water mosses and liverworts

Family: Anthomyiidae
(Example)

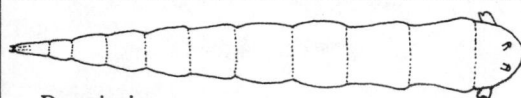

Limnophora sp.

Description
Last abdominal segments with a pair of large fleshy hooks.
Colour: Whitish or pale yellow
Length: Up to 25 mm
Genus: *Limnophora*

Occurrence
Among vegetation in streams particularly where there is a strong current

Family: Chironomidae
(Examples)

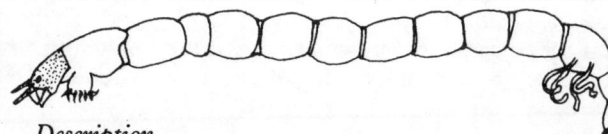

Chironomus sp.

Description
Last abdominal segment with four retractile anal gills. Second and last abdominal segments with four tubular gills on the ventral surface.
Colour: Variable. Transparent colourless to red

Length: Up to 20 mm
Genus: *Chironomus*

Occurrence
Found in all kinds of water, may be particularly abundant in the mud of organically polluted streams and rivers. Some species live among vegetation, others in mud tubes

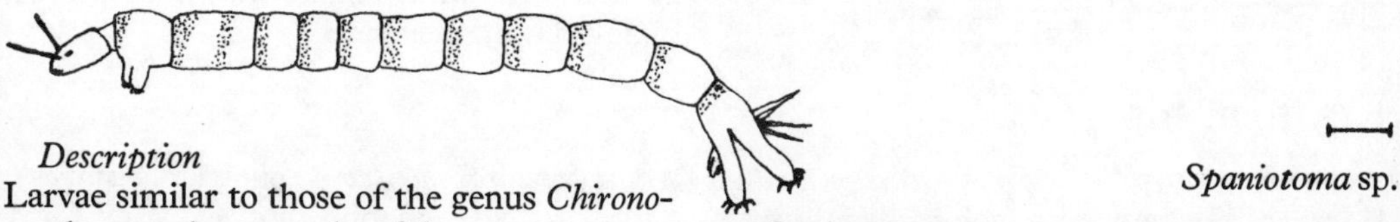

Spaniotoma sp.

Description
Larvae similar to those of the genus *Chironomus* but much more colourful.
Colour: Yellowish, greenish, bluish
Length: Up to 7 mm
Genus: *Spaniotoma (Orthocladius)*

Occurrence
In streams and rivers, either free, in gelatinous tubes or tubes with sand grains attached

Order: Trichoptera—Larvae

The larvae of the order Trichoptera (caddis flies) are common in most aquatic habitats, and many should be found in a study of streams and rivers. All the larvae are difficult to identify accurately, especially the cased caddises. Each specimen will need particularly careful study to determine the family to which it belongs, before identifying to species (if this should prove necessary).

The keys provided here are as simple as possible and attempts have been made to avoid examination of parts of the specimen which are likely to cause real difficulty. It is hoped that the diagnostic features mentioned here will be easily observed with the aid of a good stereo, binocular microscope or by using the low power of an ordinary microscope with the specimen illustrated from the side.

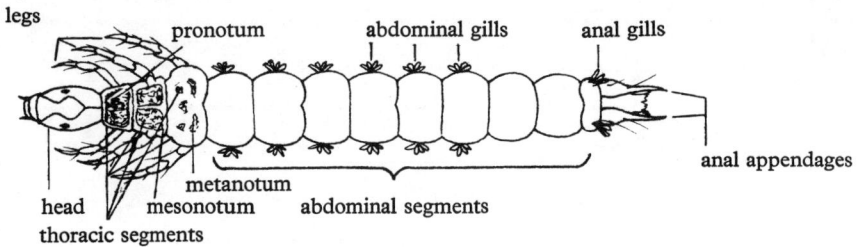

Special note:

1 Sclerotized tissue—tissue which is thickened to form a tough, hard, impermeable layer. This type of tissue may be seen on the upper surfaces (nota) of the three thoracic segments of caddis larvae. The surfaces will appear darker and tougher than the surrounding softer, fleshy, membranous tissue.

2 Head-front view—as an aid to identification of species diagrams of the head in front view are provided with the other diagrams and details of the species in this guide.

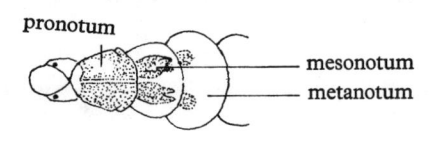

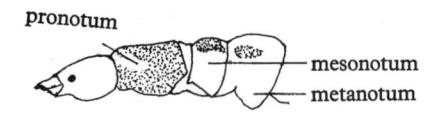

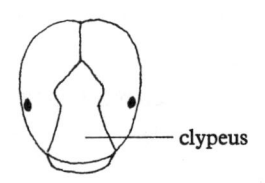

3 In order to examine the cased caddises properly the individuals will have to be removed from their cases.

Key to groups:

1 Larvae free, without a mobile case. Larvae often live in nets or galleries. Head flattened, anal claws always well-developed
   ☐ caseless caddis larvae
      see pages 53–56

2 Larvae not free (except young larval stages), enclosed in a mobile case of varied form and of mineral or vegetable material. Head rarely flattened, anal claws short
   ☐ cased caddis larvae
      see pages 56–63

1 Caseless caddis larvae

Examine carefully and determine

(a) degree of sclerotization of the pro-, meso- and metanota (dorsal surface of the three thoracic segments). Illustrated with the diagram representing the family

(b) the presence or absence of abdominal and anal gills

(c) the patterning on the head, especially the clypeus

Key to families:

1 Pro-, meso- and metanota sclerotized (Gills in bunches on the ventral surface of most abdominal segments and on two of the thoracic segments. Anal gills may be present. Abdomen covered with short, black hairs)

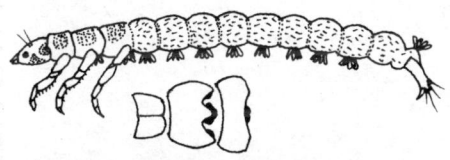

☐ Family: Hydropsychidae
see page 54

Only pronotum sclerotized
☐ 2

2 Gills present either on abdominal segments or as anal gills at end of abdomen, or both
☐ 3

Gills absent, but end of abdomen with long anal appendages (Legs short and unequal in length. Abdomen reddish with whitish lateral lines)

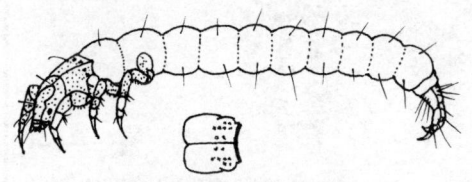

☐ Family: Polycentropidae
see page 54

3 Abdominal and anal gills present (Abdominal gills tufted structures along sides of segments 1–8. Six anal gills present. Legs almost equal in length)

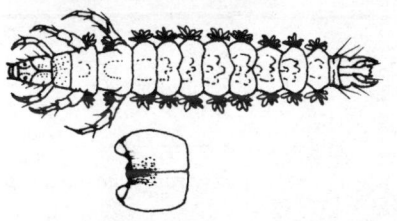

☐ Family: Rhyacophilidae
see page 55

Only anal gills present
☐ 4

4 Five short anal gills present, anal appendages well-developed (Legs unequal in length, with two spurs on the tibia)

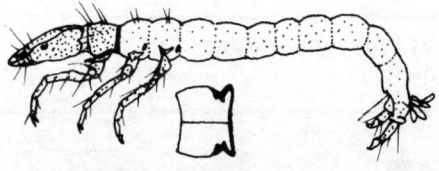

☐ Family: Philopotamidae
see page 55

Five long anal gills present (Larvae small, inhabit tunnels. Usually brightly coloured)

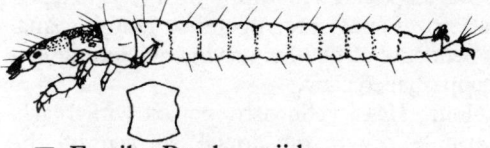

☐ Family: Psychomyiidae
see page 56

Family: Hydropsychidae
(Examples)

Hydropsyche instabilis

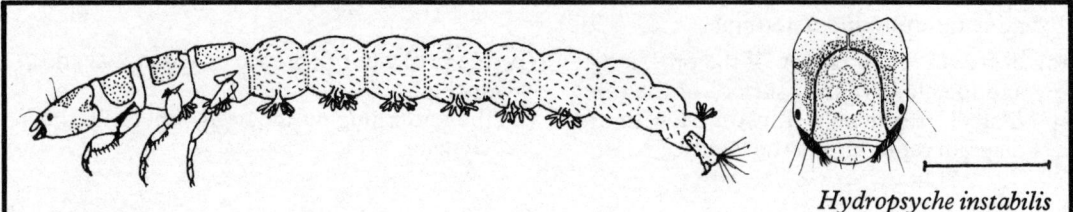

Description
Gills with many branches present on meso- and meta-thoracic segments and on abdominal segments 1–6. Four anal gills present. Colour: Head yellow with dark brown markings. Clypeus marked with four yellow areas. Thorax greyish-brown

Length: Up to 17 mm
Hydropsyche instabilis

Occurrence
In elongated shelters attached to the substratum, a net is stretched from one end of the shelter. May be abundant in streams and small rivers where the current is swift

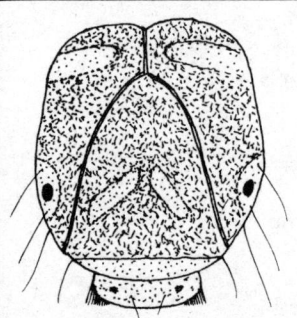

Hydropsyche angustipennis

Description
Similar to *Hydropsyche instabilis* but with very branched gills on the ventral surface of abdominal segments 1–7. Head with a ∧-shaped light patch on the clypeus.

Colour: Head dark yellowish-brown, body yellowish or greyish-brown
Length: Up to 17 mm
Hydropsyche angustipennis

Occurrence
Live in rough, elongated silken shelters attached to the underside of stones in streams and rivers

Family: Polycentropidae
(Example)

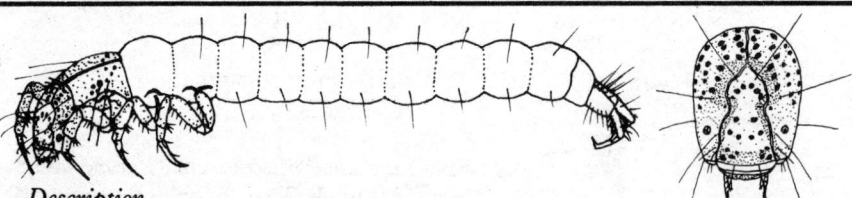

Description
Head elliptical. Pronotum with a W-shaped pattern of spots near posterior margin. Anal appendages of two long segments ending in a scoop-shaped claw.
Colour: Head yellowish spotted with black. Variable colour on dorsal surface of the abdomen—brownish-red, brownish-yellow, pale green. Ventral surface whitish

Polycentropus flavomaculatus

Length: Up to 14 mm
Polycentropus flavomaculatus

Occurrence
Common, found wherever there is moving freshwater. Constructs small silken nets under stones, etc

Family: Rhyacophilidae
(Example)

Rhyacophila dorsalis

Description

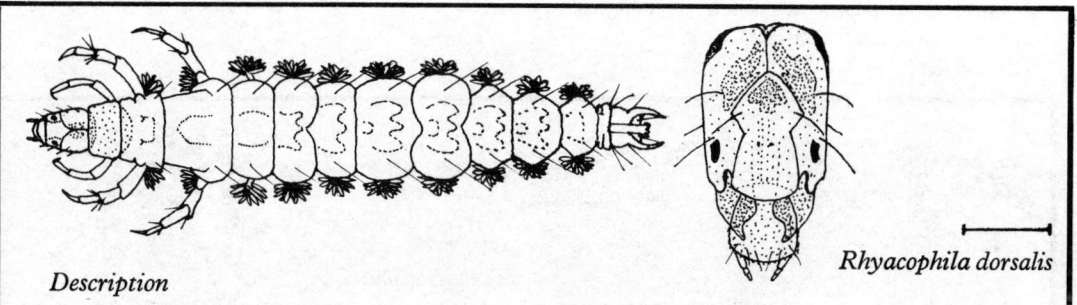

Head long. Posterior and lateral margins of the pronotum convex, anterior margin concave. Tuft of filamentous gills at bases of the second and third pairs of legs and abdominal segments 1–8. Ninth abdominal segment with a heavy sclerotized band.

Colour: Head and thorax cream, apart from sclerotized areas, abdomen greenish

Length: Up to 12 mm
Rhyacophila dorsalis

Occurrence
Characteristic of fast flowing streams, typically found under stones

Family: Philopotamidae
(Example)

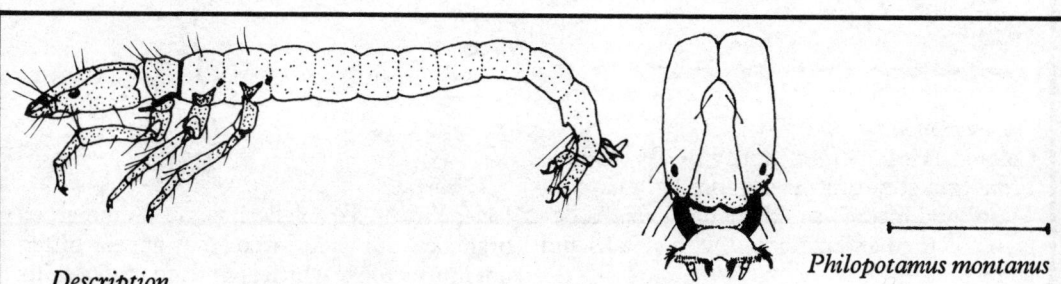

Philopotamus montanus

Description

Head long, Tarsal claws of all the legs with a spine, well seen in the second and third pairs of legs. Anal gills present on the dorsal surface of the last abdominal segment.

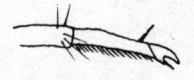

Colour: Head and thorax chestnut-brown with darker margins, abdomen white

Length: Up to 22 mm
Philopotamus montanus

Occurrence
In streams which are well-aerated and rocky. Common in rapid hillside streams in the west of Britain

Family: Psychomyiidae

Note: Colour can be important in recognizing species. Construct tunnels.
(Examples)

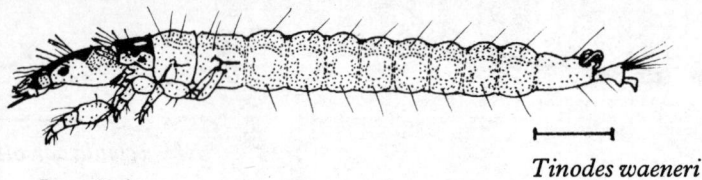

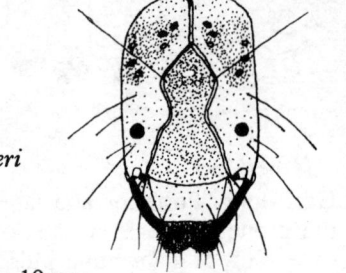

Tinodes waeneri

Description
Head oval and long, eyes in forward position. Legs small in relation to body size. Ninth segment of abdomen much narrower than first eight segments. Five anal gills present.
Colour: Pronotum dark brown with four large creamy areas. Head yellowish-green, clypeus brownish-grey. Meso- and metanota chestnut-brown, ventral regions apple green. Abdomen with dorsal surface bright green

Length: Up to 10 mm
Tinodes waeneri

Occurrence
May be found in slow rivers. Constructs a tunnel of sand grains, mud and vegetable debris

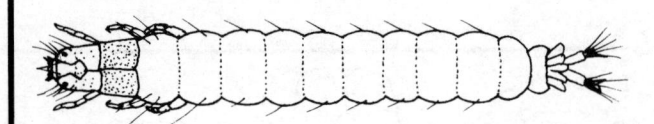

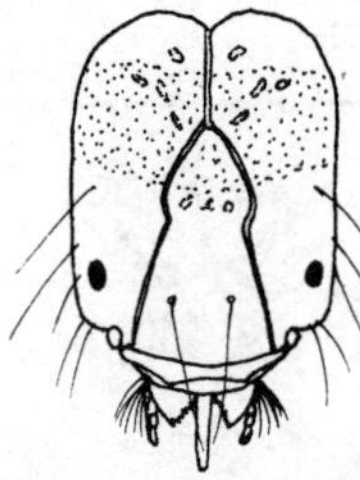

Description
Colour: Head—front yellow, rest brown. Pronotum grey-brown with four light patches Meso- and metanota a pinkish colour. Abdomen coloured and patterned with streaks and spots
Length: Up to 11 mm
Lype phaeopa

Lype phaeopa

Occurrence
Lives in fixed galleries on submerged branches. Surface of wood may appear bitten out into a groove which is covered with a skin-like secretion. Not easily found but may be common

2 Cased caddis larvae

Examine carefully and determine
(a) degree of sclerotization of the pro-, meso- and metanota (dorsal surface of the three thoracic segments) illustrated with the diagram representing the family
(b) the presence or absence of abdominal and anal gills
(c) the patterning on the head, especially the clypeus
(d) the shape and type of case

Key to families:

1 Pro-, meso- and metanota completely sclerotized except for the margins or only the pronotum sclerotized and the other nota membranous
☐ 2

Pronotum completely sclerotized, but the meso- and metanota with varying degrees of sclerotization
☐ 3

2 Pro-, meso- and metanota completely sclerotized (Abdominal segments may have sclerotized patches. Gills absent. Larvae small. Case vase or seed-shaped, made of sand grains)

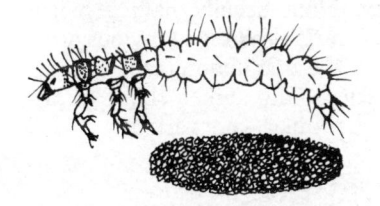

☐ Family: Hydroptilidae
see page 58

Only pronotum sclerotized, meso- and metanota membranous (Strong prosternal horns (protuberances from the first abdominal segment). Abdomen broad and fat. Simple thread-like gills on abdominal segments. Case cylindrical, of vegetable material)

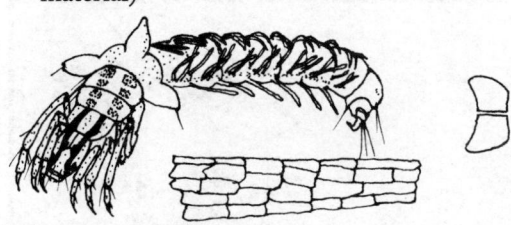

☐ Family: Phryganeidae
see page 59

3 Mesonotum completely sclerotized in the form of two plates
☐ 4

Mesonotum only partially sclerotized
☐ 6

4 Metanotum of small sclerites (Larvae fat and body soft, large dorsal hump on first abdominal segment. Numerous gills usually present on abdominal segments. Case usually cylindrical of sand grains, stones or vegetable material)

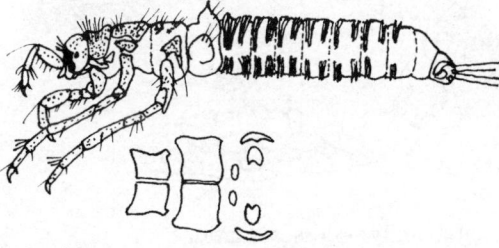

☐ Family: Limnephilidae
see page 59

Metanotum partially sclerotized or membranous
☐ 5

5 Metanotum with four sclerotized patches (Large dorsal hump on first abdominal segment. Thread-like gills on abdominal segments 2–7. Case tubular, of sand grains, smooth in texture)

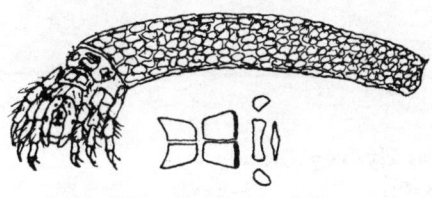

☐ Family: Odonticeridae
see page 60

Metanotum membranous (Body slender and cylindrical. Legs unequal in length, third pair longest. Case narrow and conical, of sand grains or vegetable material)

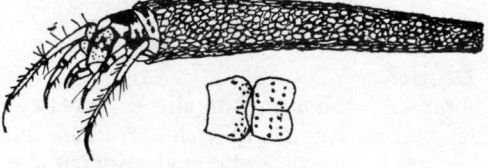

☐ Family: Leptoceridae
see page 60

6 Mesonotum partially sclerotized, sclerotized areas extending into posterior projections. Metanotum with a pair of small sclerotized patches (Abdominal gills absent, six anal gills present. Case of stone chips, pebbles, etc)

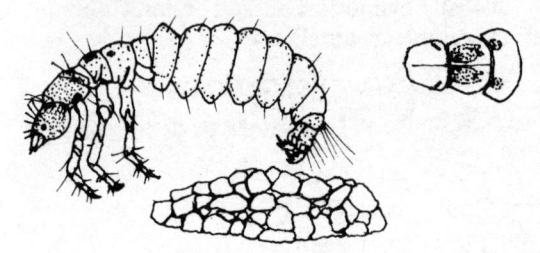

☐ Family: Glossosomatidae
see page 60

Mesonotum sclerotized but without posterior projections
☐ 7

7 Mesonotum of three plates, metanotum membranous (One dorsal and two lateral protuberances on first abdominal segment. Thread-like gills on abdominal segments. Case shield-shaped, broad and flat, of sand grains)

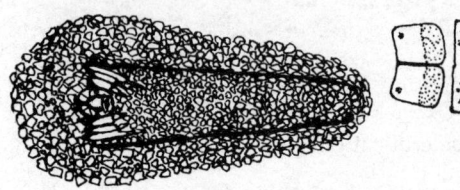

☐ Family: Molannidae
see page 61

Mesonotum with sclerotized patches, metanotum with sclerotized patches or membranous. Nota usually have a hairy appearance (A variable family. Obvious protuberances on the first abdominal segment of some species. Gills either single or in bunches. Case tubular, very variable)

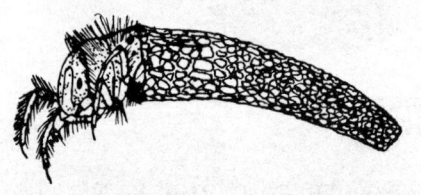

☐ Family: Sericostomatidae
see pages 61–63

Family: Hydroptilidae
(Example)

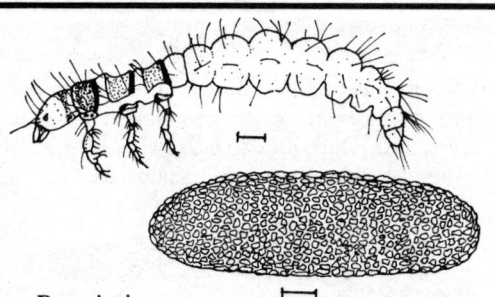

Hydroptila sparsa

Description
Thorax and abdomen laterally compressed. First pair of legs slightly shorter than the other pairs. Fourth abdominal segment the largest. Case of very fine sand grains.
Colour: Head and notal regions greyish-brown

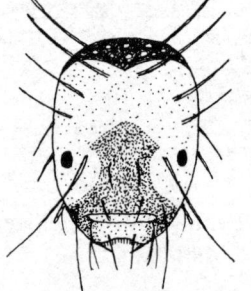

Length: Larva up to 4 mm, case up to 5 mm
Hydroptila sparsa

Occurrence
Widely distributed and generally abundant in streams and rivers

Family: Phryganeidae
(Example)

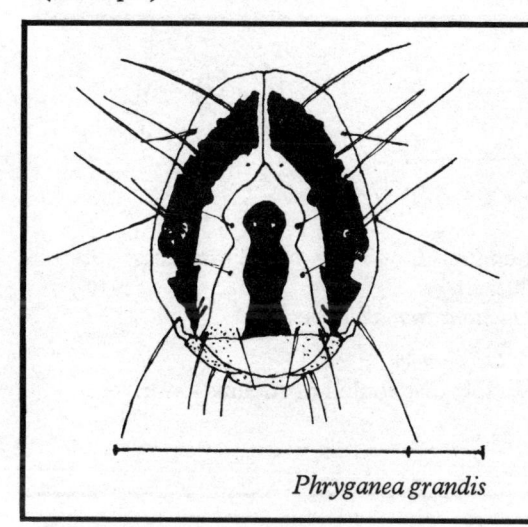

Phryganea grandis

Description
First abdominal segment with a large pointed process directed posteriorly, and a pair of blunt lateral processes which curve forwards. Remaining abdominal segments with filamentous gills. Case of regularly arranged leaf fragments.
Colour: Head bright yellow with a distinct ∧-shaped mark. Pronotum yellow
Length: Larva up to 40 mm, case up to 50 mm
Phryganea grandis

Occurrence
Commonly found in ponds, lakes and slow-flowing streams and rivers

Family: Limnephilidae
(Examples)

Description
Head elliptical covered with minute spines and numerous long hairs. First pair of legs shortest with dark spots on tibia. Dorsal and lateral protuberances on abdominal segments. Filamentous gills on thoracic and abdominal segments. Case of vegetable material and debris.

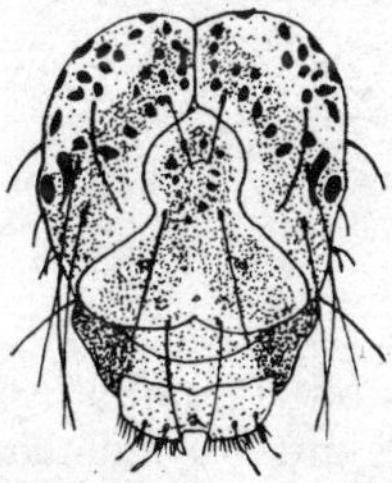

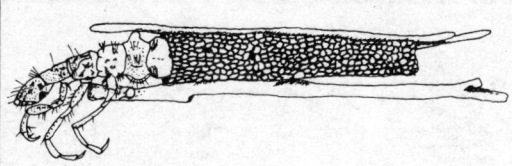

Limnephilus lunatus

Colour: Head yellowish brown, pronotum dark brown, abdomen white
Length: Larva up to 17 mm, case up to 23 mm
Limnephilus lunatus

Occurrence
In streams and rivers but may be found in most waters where there are strong growths of aquatic plants

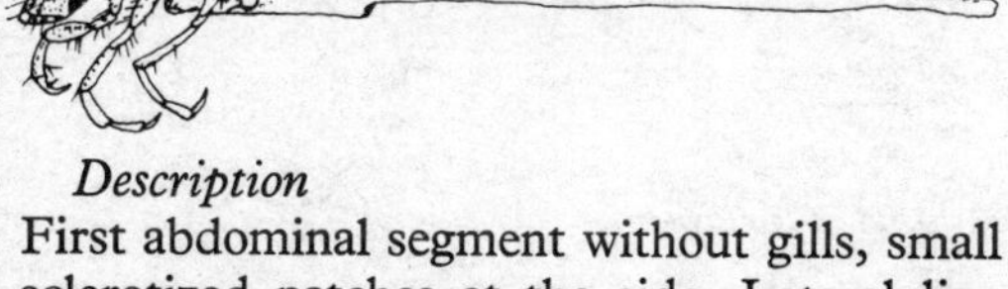

Description
First abdominal segment without gills, small sclerotized patches at the side. Lateral line present. Thread-like gills on abdominal segments 2–7. Case conical, of sand grains with twigs positioned longitudinally.
Colour: Head yellowish-brown with dark markings. T-shaped mark on clypeus. Thoracic segments with dark markings

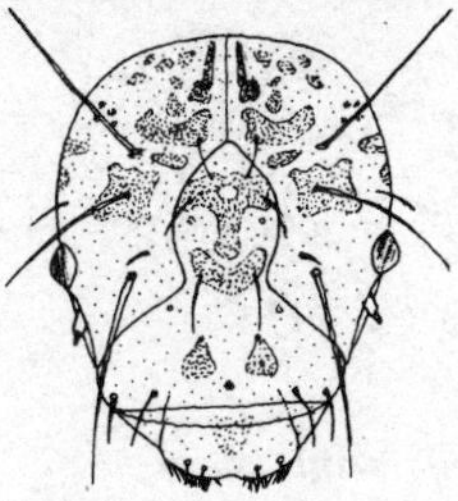

Anabolia nervosa

Length: Larva up to 18 mm, case up to 26 mm
Anabolia nervosa

Occurrence
Widely distributed in running and still water

Family: Odontoceridae
(Example)

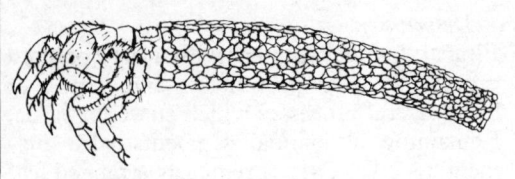

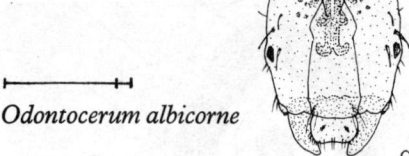

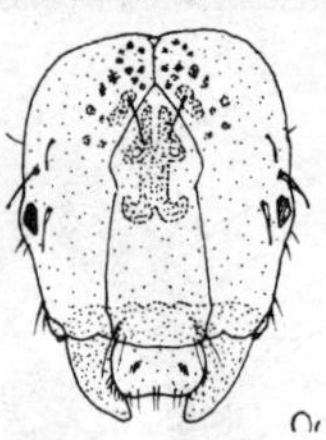

Odontocerum albicorne

Description
Clypeus with an anchor-shaped mark. Stump-like dorsal sucker on first abdominal segment. Gills thread-like in circlets near anterior margin of abdominal segments 2–7. Case slightly curved, tapering to the posterior end, made of sand grains.
Colour: Head light brown with light and dark spots

Length: Larva up to 18 mm, case up to 20 mm
Odontocerum albicorne

Occurrence
Widely distributed in streams and rivers

Family: Leptoceridae
(Example)

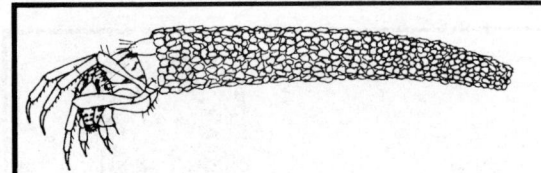

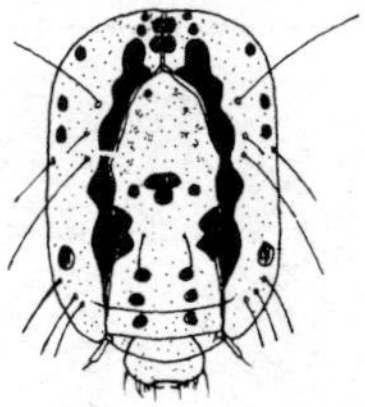

Athripsodes aterrimus

Description
Gills on abdominal segments 1–3, consist of small tufts of filaments on a common stalk. Case curved and tapering posteriorly, made of sand grains.
Colour: Black spots on sides of pronotum, abdomen creamy white

Length: Larva up to 11 mm, case up to 18 mm
Athripsodes aterrimus

Occurrence
Found in slow-running streams and rivers where the substratum is sandy or gravelly

Family: Glossosomatidae
(Example)

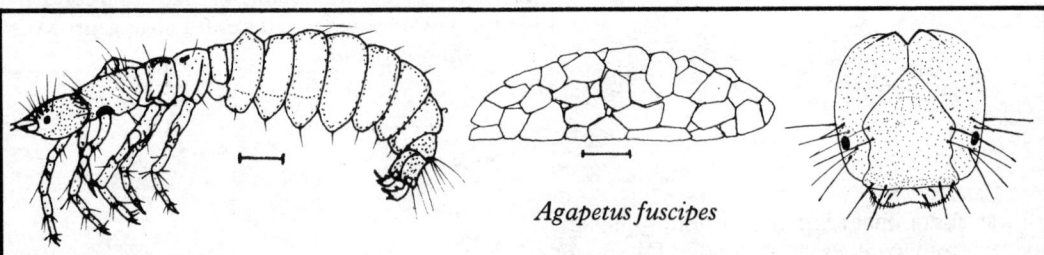

Agapetus fuscipes

Description
Head oval, clypeus wide. Legs roughly equal in length. Abdomen curved ventrally. Case of stones, small pebbles, sand grains. Hemispherical dorsally, flattened ventrally.
Colour: Head and pronotum dark brown

Length: Larva up to 6 mm, case up to 8 mm
Agapetus fuscipes

Occurrence
Widely distributed and common in shallow, swiftly flowing streams and rivers with a stony substratum

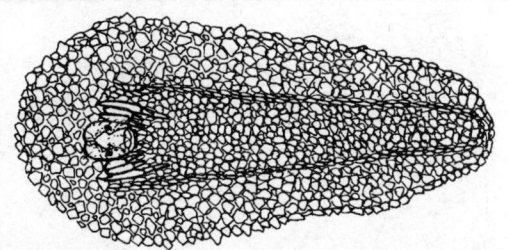

Family: Molannidae
(Example)

Description
Thread-like gills on abdominal segments 1–8, small groups of usually three gills. Case: shield-shaped of sand grains with a central conical tube.
Colour: Head light yellowish-brown with a dark band running longitudinally. Abdomen greyish-white

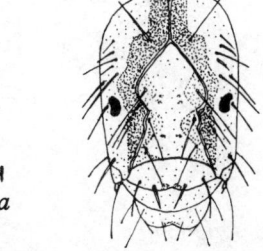

Molanna augustata

Length: Larva up to 17 mm, case up to 26 mm
Molanna augustata

Occurrence
Found in slow-running streams and rivers where the substratum is sandy or gravelly

Family: Sericostomatidae

The members of this family show wide variations and for this reason the four sub-families are considered:
Examine carefully and determine
(a) colour of the head and whether retractile into the prothorax
(b) degree of sclerotization on the nota
(c) type of gills
(d) construction of the case

Key to the sub-families:

1 Head dark brown with pattern of pale spots. Head retractile into prothorax
☐ 2

Head not dark brown, but retractile into prothorax
☐ 3

2 Pro- and mesonotum dark brown, anterior half of pronotum with obvious dark hairs. Metanotum without sclerotized patches (Gills either simple and thread-like or in bunches. Case smooth, curved and tapered, of sand grains)
☐ Sub-family: Sericostomatinae
see page 62

Anterior angles of pronotum project forward. Mesonotum with two large sclerotized patches, metanotum with 3 or 4 sclerotized patches (2 or 3 thread-like gills on dorsal and ventral surfaces of segments 2–7. Case tubular of sand grains with larger stones attached)
☐ Sub-family: Goeridae
see page 62

3 Head golden-brown with dark areas on anterior and posterior margins. Posterior margin of pronotum heavily sclerotized, mesonotum lightly sclerotized, metanotum with 4 sclerotized patches, arranged in a crescent-shape (Gills tufted on dorsal surface near posterior margin, become progressively reduced in size towards the seventh segment. Case circular in section, in older larvae constructed of body secretions, base attached to objects)
☐ Sub-family: Brachycentrinae
see page 63

Head appears reddish to the naked eye, chestnut-brown on close examination. Mesonotum partially or completely sclerotized, metanotum with three or less sclerotized patches (Pairs of single gills on dorsal and ventral surfaces of segments 2–6. Case of vegetable material, square in cross-section, tapering away from the head end)
☐ Sub-family: Lepidostomatinae
see page 63

Sub-family: Sericostomatinae
(Example)

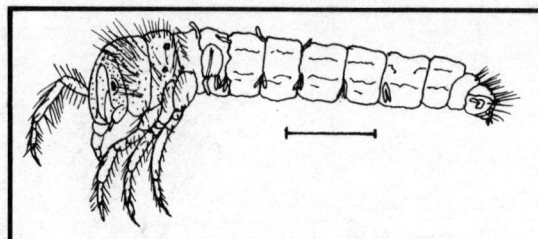

Description
Head retracted into prothorax. Light pear-shaped marks on head. Flattened disc-like protuberances on first abdominal segment. Gills on segments 1–6, either simple or in small bunches. Case: delicate mosaic of sand grains
Colour: Head dark brown. Pronotum greyish-brown; abdomen cream with red patches

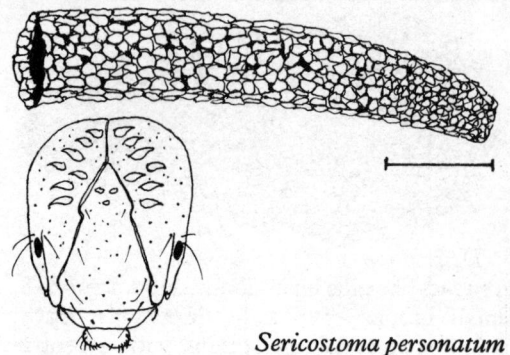

Sericostoma personatum

Length: Larva up to 12 mm, case up to 14 mm
Sericostoma personatum

Occurrence
Fast streams and rivers, also in lakes where streams enter

Sub-family: Goerinae
(Example)

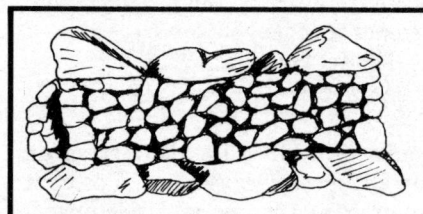

Description
Head covered with pustules and recessed into angular extensions of the prothorax. Two quadrangular sclerotized patches on the mesonotum and patches on the metanotum. Gills thread-like on segments 2–7.
Case: Cylindrical part made of sand grains, with wing-like extensions of stone fragments or small pebbles.

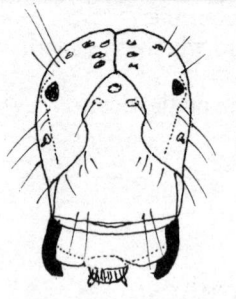

Goera pilosa

Length: Larva up to 13 mm, case up to 15 mm
Goera pilosa

Occurrence
Generally common in fast flowing streams

Sub-family: Brachycentrinae
(Example)

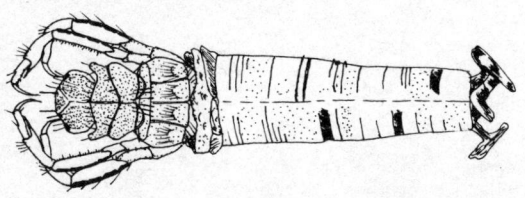

Description
First pair of legs short and very broad. Second and third pairs much larger. Tufted gills on dorsal surface of abdominal segments 2–7, becoming progressively reduced in size towards the seventh segment.
Case: of young larvae of vegetable debris, in older larvae of body secretions, attached to plants and stones.

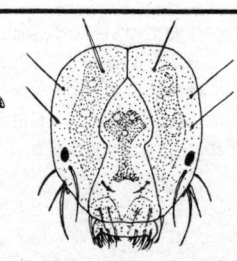

Brachycentrus subnubilus

Colour: Head golden-brown with dark anterior and posterior margins. Longitudinal bands on sides of head
Length: Larva up to 12 mm, case up to 12 mm
Brachycentrus subnubilus

Occurrence
Widely distributed in slow flowing streams and rivers

Sub-family: Lepidostomatinae
(Example)

Description
Lateral protuberances on first abdominal segment. Pairs of single thread-like gills on abdominal segments 2–6. A pair of triangular protuberances on abdominal segment 8.
Case: square in section of sand grains or vegetable material such as dead leaves or rotting wood.
Colour: Head chestnut or reddish
Length: Larva up to 7 mm, case up to 9 mm
Crunoecia irrorata

Crunoecia irrorata

Occurrence
Widely distributed but rarely abundant in small fast streams

Order: Coleoptera—Larvae

No attempt is made here to separate out the various Coleoptera larvae into species. The descriptions given below serve only to distinguish between those families of larvae that are likely to be encountered in streams and rivers.

Examine carefully and determine
(a) the presence or absence of obvious lateral and dorsal projections from the abdominal segments
(b) the types of tail processes

Key to families

1 Larvae with obvious long projections arising from the abdominal segments
   ☐ 2

   Larvae without any obvious abdominal projections
   ☐ 3

2 Body rather robust, elongate with long lateral projections (gills) on the abdominal segments. Last abdominal segment with two pairs of curved hooks

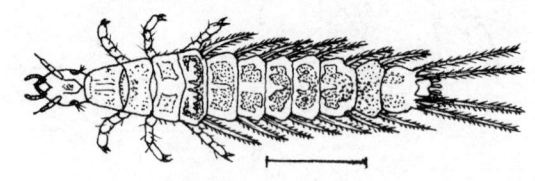

   ☐ Family: Gyrinidae

   Body long and thin, with spine- or thread-like projections from the abdominal segments. Last abdominal segment ends in a single or double tail-like process

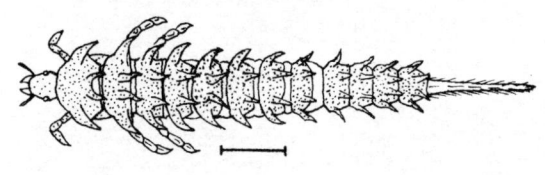

   ☐ Family: Haliplidae

3 Body cylindrical and tapering towards the end of the abdomen. Two tail-like processes extend from the end of the abdomen

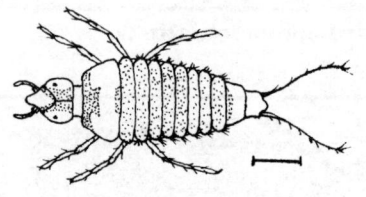

   ☐ Family: Dytiscidae

   Body broad and flat. Abdomen terminates in a feathery process

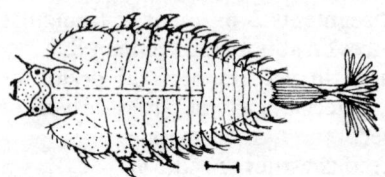

   ☐ Family: Helmidae (Elmidae)

Class: Insecta—Pupae

The term pupa refers to the resting stage of holometabolous insects (insects with a complex metamorphosis—wings develop internally and the immature stages are larvae which differ from the adults in structure and habits). During this stage the larval body and its internal organs are remodelled to adapt them to the requirements of the adult. It should be noted that although the pupal stage is often regarded as one of quiescence, certain pupae are capable of active movements.

It is beyond the scope of this book to deal with the pupal stages in any more than a general way. There is no attempt to give any detailed account or to identify the pupae to species.

Key to groups:

1 Pupae encased in a simple protective cuticle or cocoon, with few obvious appendages
 ☐ Diptera pupae

 (a) Cocoon elongate with projecting thoracic respiratory horns or plates. Found in mud or other soft deposits
  ☐ Family: Tipulidae
 (b) Cocoon elongate, thoracic respiratory appendages connected to a large median, anterior cavity. Terminal segment with six stout projections. Abdominal segments banded. Found in mud or other soft deposits
  ☐ Family: Tabanidae
 (c) Cocoon small and pocket-like, attached to stones, etc. Long respiratory filaments extend from the cocoon

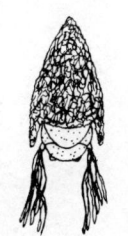

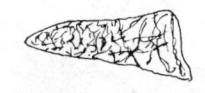

  ☐ Family: Simuliidae

 (d) Pupae rather elongated and somewhat comma-shaped. May be active and float at the surface or remain in the bottom mud

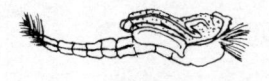

  ☐ Family: Chironomidae

2 Pupae usually soft and pale in colour, may possess a thin, soft cuticle. Pupae usually found on land near water. Only likely to be encountered if searched for
 ☐ Coleoptera pupae

3 Pupae encased in pupal shelters of various designs, e.g. caddis larvae cases (Trichoptera) with plugging material
 ☐ Trichoptera pupae

 (a) Silken wall constructed across ends of existing larval cases, sometimes strengthened by plant fragments and stones. Pupae lie free in the case

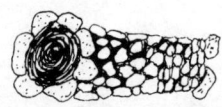

  ☐ Cased caddises

 (b) Pupal shelters—oval, cavern-like structures constructed of small stones, sand grains, etc. Pupae enclosed in brownish cocoons within
  ☐ Caseless caddises

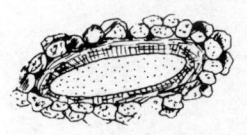

 e.g. *Rhyacophilus dorsalis*—attached to stones—removed showing pupa from below.

Class: Arachnida

The common representatives of this class are members of a group called the Hydracarina (water mites). Some 250 species are found in aquatic habitats in Britain and these are grouped into two families. The majority of species are found among vegetation or attached to other aquatic animals, particularly insects, in slow-flowing streams, rivers, canals, ponds and lakes. Only a few are found in the faster flowing areas of streams and rivers.

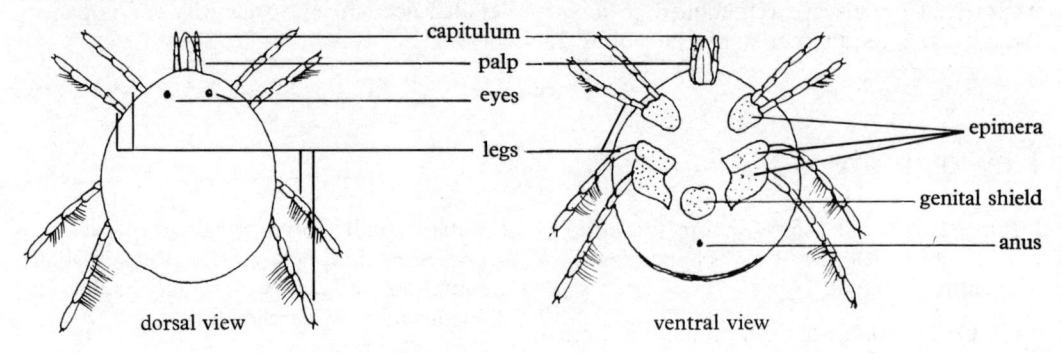

dorsal view — capitulum, palp, eyes, legs

ventral view — epimera, genital shield, anus

Key to families:

1 Skin usually soft but may have hard plates on the back. Two eyes present and enclosed in capsules. Sometimes a third eye is present between the other pair. Epimera of legs, on ventral surface, in four groups, a pair of legs arising from each. Genital area always near the epimera.
Colour: Always some shade of red
Length: Up to 8 mm

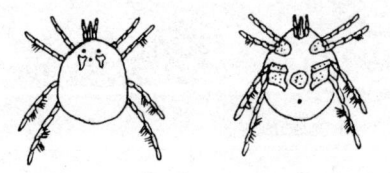

☐ Family: Limnocharidae

2 Skin may be soft or heavily sclerotized. Two eyes present but not enclosed in capsules. A third eye never present. Epimera cover a large area of the ventral surface, and the pairs of epimera are usually close together. The genital area is usually well behind the epimera.
Colour: Very variable, red, brown, blue, yellow, green or mixtures
Length: Up to 4 mm

☐ Family: Hygrobatidae

Phylum: Mollusca

Two classes of this phylum are represented in fresh water.

1 One-shelled, asymmetrical molluscs. Head well-developed and bears a mouth, one or two pairs of retractile tentacles. Moves on a large foot

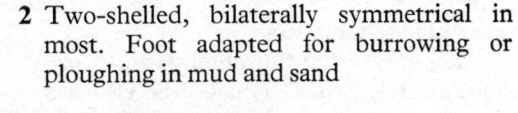

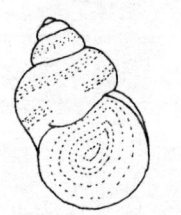

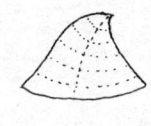

☐ Class: Gastropoda
see pages 68–74

2 Two-shelled, bilaterally symmetrical in most. Foot adapted for burrowing or ploughing in mud and sand

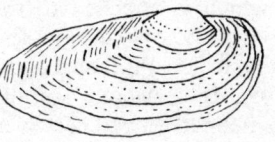

☐ Class: Lamellibranchiata
see pages 75–78

Measurement

In this section measurements are given for each member of the two classes considered. Height and width are considered in the members of the class Gastropoda whilst height, width and length are considered in the class Lamellibranchiata. Measurements can be made with the aid of callipers as follows:

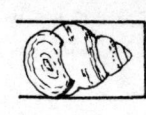

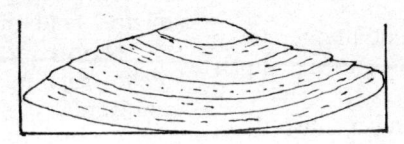

height   width   length

Class: Gastropoda

Two orders of gastropod molluscs occur in streams and rivers, namely the Prosobranchiata and the Pulmonata.

Members of the Prosobranchiata possess a spirally coiled shell. This can be closed by a flap or operculum which is attached to the hind part of the foot. These snails possess external gills. In the order Pulmonata the shell is more variable. It is usually spirally coiled, but may be helmet-shaped, or even very much reduced in size. There is no operculum to close the mouth of the shell.

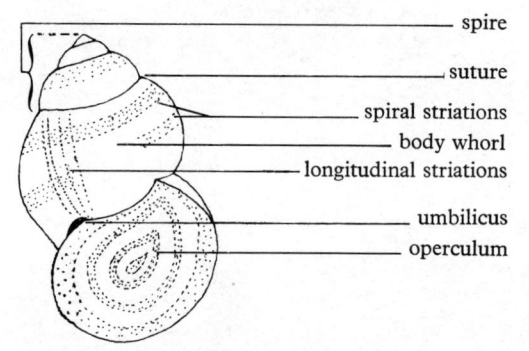

Prosobranch gastropod

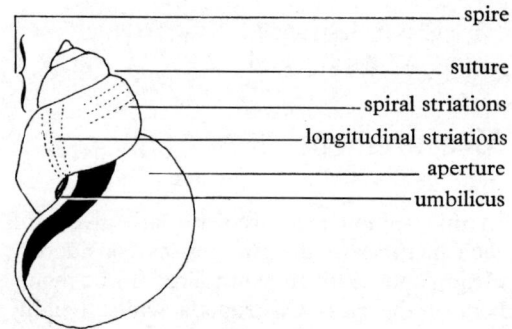

Pulmonate gastropod

Examine the snail carefully and determine:

(a) the presence or absence of an operculum
(b) the number of tentacles—either one pair or two pairs
(c) whether the shell is dextral or sinistral

To determine whether a shell is dextral or sinistral hold the shell upright and look at the aperture. If the aperture is on the right side the shell is dextral, if on the left the shell is sinistral.

Key to order:

1 Snails with an operculum and one pair of tentacles

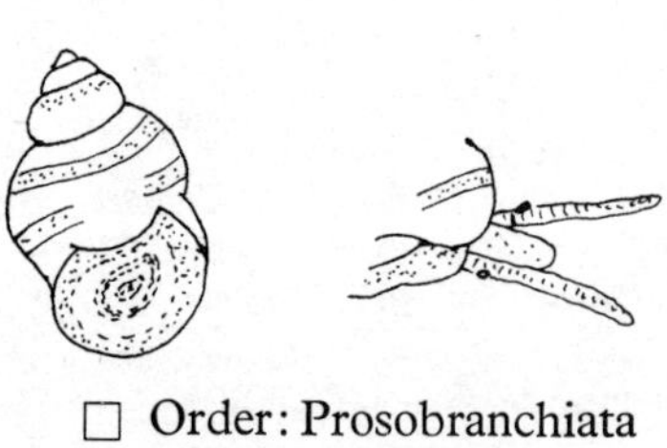

☐ Order: Prosobranchiata
see pages 69–71

2 Snails without an operculum and two pairs of tentacles

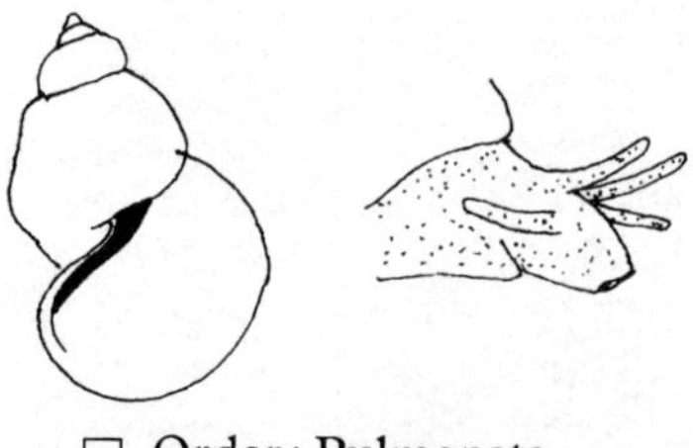

☐ Order: Pulmonata
see pages 72–74

1 Order: Prosobranchiata
All shells dextral.

Examine carefully and determine: (a) shape of shell
(b) height of shell

Key to families:

1 Shell half-moon shaped, spire very low

☐ Family: Neritidae
see page 69

Shell unlike above
☐ 2

2 Shell flat and disc-like, squat

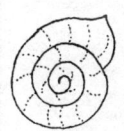

☐ Family: Valvatidae
see page 70

Shell not flat and disc-like
☐ 3

3 Shell conical and globe-like, large (up to 40 mm high). Banded, apex pointed

☐ Family: Viviparidae
see page 70

Shell cylindrical or spire-shaped. Most less than 10 mm high

☐ Family: Hydrobiidae
see page 71

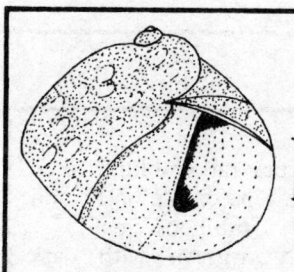

Family: Neritidae
(Example)

Theodoxus fluviatilis

Description
Shell thick-walled, spire very low. Mouth of shell semi-circular. Operculum spirally grooved. Umbilicus present.

Colour: Variable, whitish, yellowish or pale grey
Height: up to 8 mm, width: up to 12 mm
Theodoxus fluviatilis

Occurrence
In hardwater streams and rivers, particularly in the more rapid parts where the substratum is composed of stones, pebble and gravel

Family: Valvatidae
(Examples)

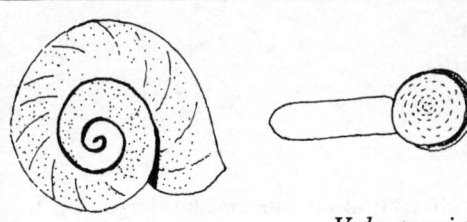

Valvata cristata

Description
Shell disc-like, thin and glossy, with deep sutures. Umbilicus wide. The snail possesses one thread-like and one feathery gill.

gills

Foot is forked.

Height: up to 2 mm, width: up to 5 mm
Valvata cristata

Occurrence
Fairly common in slow-running water where there is plenty of mud round vegetation. Not recorded in Cornwall or the Scottish Highlands

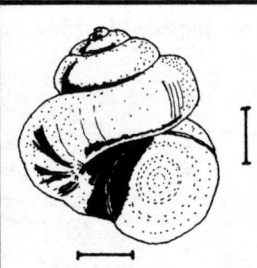

Valvata piscinalis

Description
Shell larger than *Valvata cristata*. Rather globular with a blunt conical spire. Sutures deep. Mouth of shell almost circular, operculum spirally grooved. Umbilicus narrow. One thread-like and one feathery gill present.

Colour: Yellowish or greenish, shiny
Height: up to 7 mm, width: up to 7 mm
Valvata piscinalis

Occurrence
Common in most running waters where the current is not strong

Family: Viviparidae
(Example)

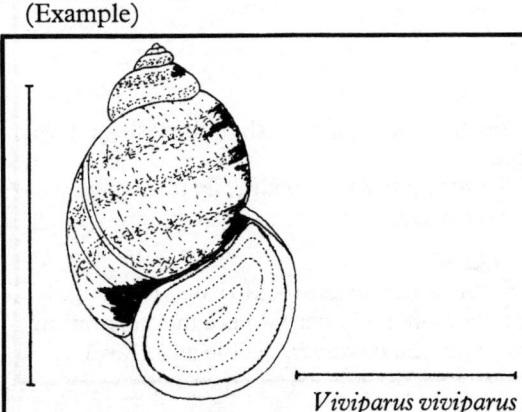

Viviparus viviparus

Description
Shell large, conical, apex rather blunt. Shell mouth obliquely oval. Operculum horny and concentrically ringed. Umbilicus small.
Colour: Usually brown-green with dark bands (bands may be absent)
Height: up to 40 mm, width: up to 30 mm
Viviparus viviparus

Occurrence
Common in streams and rivers with much vegetation, in England and Wales to Yorkshire. Not found in the west of England

Family: Hydrobiidae
(Examples)

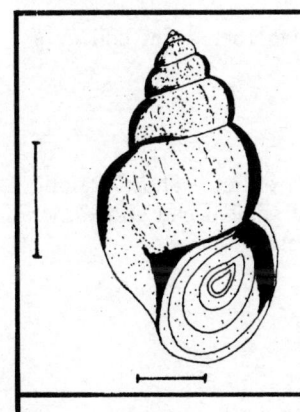

Bithynia tentaculata

Description
Shell conical, base rounded, spire pointed and sutures shallow. 5–6 whorls. Shell mouth oval, operculum concentrically grooved, limy. Umbilicus very small.
Colour: Yellowish
Height: up to 15 mm, width: up to 9 mm
Bithynia tentaculata

Occurrence
Limited in distribution. Prefers hardwaters of slow-running streams and rivers

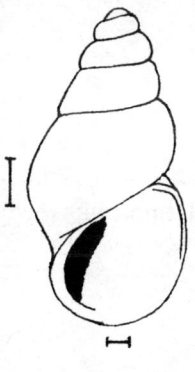

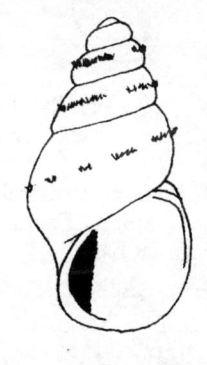

Potamopyrgus jenkinsi

Description
Shell pointed, spire-like. May have a spiral keel or row of spines. $5\frac{1}{2}$ whorls. Shell mouth pear-shaped. Umbilicus closed. Body of snail grey, tentacles slightly tapered.
Colour: Yellowish, crusted with black, may appear quite black
Height: up to 6 mm, width: up to 3 mm
Potamopyrgus jenkinsi

Occurrence
Common in all types of running water. May be found in vast numbers on stones, vegetation or mud

Order: Pulmonata

All the spirally coiled snails in this order are dextral with the exception of the family Physidae where the shell is sinistral.

Key to families:

1 Shell without coiling

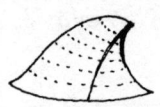

☐ Family: Ancylidae
see page 72

Shell coiled
☐ 2

2 Shell flat and disc-like, coiled in one plane

☐ Family: Planorbidae
see page 73

Shell globe-like, not coiled in one plane
☐ 3

Examine carefully and determine
(a) shape of shell, whether spirally coiled or not
(b) type of coiling, whether a flat coil or a spiral coil.

3 Shell dextral, with either a short or long spire. Tentacles of snail broad and rather triangular, eyes at base

☐ Family: Limnaeidae
see pages 73–74

Shell sinistral, with either a short or long spire. Tentacles of snail thin, with eyes at base

☐ Family: Physidae
see page 74

Family: Ancylidae

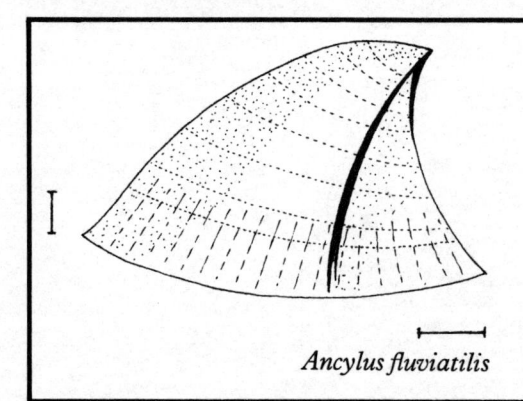

Ancylus fluviatilis

Description
Shell thin-walled, apex hooked to the right. Mouth of shell oval.
Colour: Grey to black
Height: up to 5 mm, length: up to 9 mm, width: up to 7 mm
Ancylus fluviatilis

Occurrence
Common wherever there is a hard surface, particularly in swift flowing water of streams and rivers

Family: Planorbiidae
(Examples)

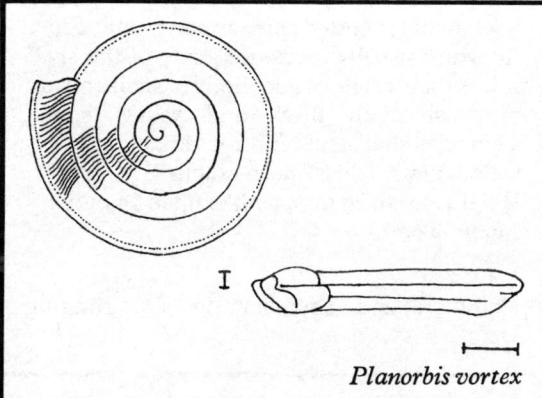

Planorbis vortex

Description
Shell small and thin. Edge with a sharp keel towards the upper spire-side of shell. Spire flat or slightly sunken, umbilicus broad and shallow.
Colour: Pale horn colour
Height: up to 2 mm, width: up to 7 mm
Planorbis vortex

Occurrence
Common in streams and rivers with plenty of vegetation. Prefers hard water

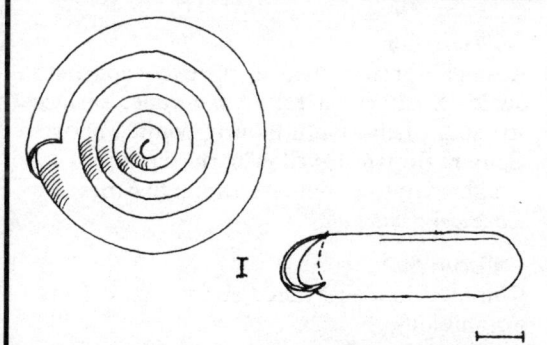

Planorbis contortus

Description
Shell small, strong-walled, thick disc. Mouth crescent-shaped. Spire deeply sunken, umbilicus slight. Shell surface finely ribbed across the whorls.
Colour: Yellowish-brown
Height: up to 2 mm, width up to 6 mm
Planorbis contortus

Occurrence
Common and widespread in flowing water of streams and rivers

Family: Lymnaeidae
(Examples)

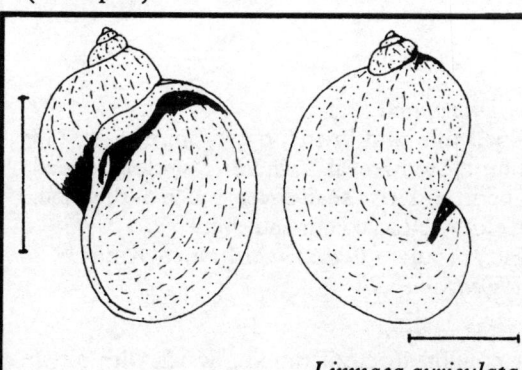

Limnaea auriculata

Description
Shell ear-shaped, broad, shell thin. Spire a short, sharp cone, pointing to the side. Aperture ear-shaped, margins usually turned outwards.
Colour: Horn coloured, brownish to whitish
Height: up to 20 mm, width: up to 14 mm
Limnaea auriculata

Occurrence
Found in slow-running streams, particularly in hard water

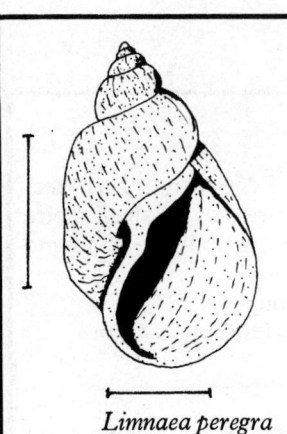

Limnaea peregra

Description
Shell ovoid, pointed spire a short, blunt cone. Aperture usually meets the body of the shell at less than a right-angle. Body of snail grey or brownish with black and yellow spots. Tentacles short and blunt. A variable species.
Colour: Pale to dark horn-colour
Height: up to 20 mm, width: up to 14 mm
Limnaea peregra

Occurrence
Common and abundant in slow-running water

Limnaea truncatula

Description
A small species, shell elongated, conical to ovoid. Shell has a turreted outline, surface striated. Shell mouth bluntly pointed above.
Colour: Brownish-yellow horn colour
Height: Up to 12 mm, width: up to 6 mm
Limnaea truncatula

Occurrence
Common in well-aerated streams and rivers, amphibious

Family: Physidae
(Example)

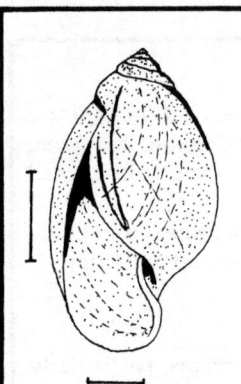

Physa fontinalis

Description
Shell thin and brittle, ovate, spire short and blunt. Mouth oval. 4 whorls. Mantle extended and in an active snail covers much of the shell.
Colour: Horn colour and shiny
Height: up to 12 mm, width: up to 8 mm
Physa fontinalis

Occurrence
Common in clean running water with much vegetation

Class: Lamellibranchiata (bivalves)

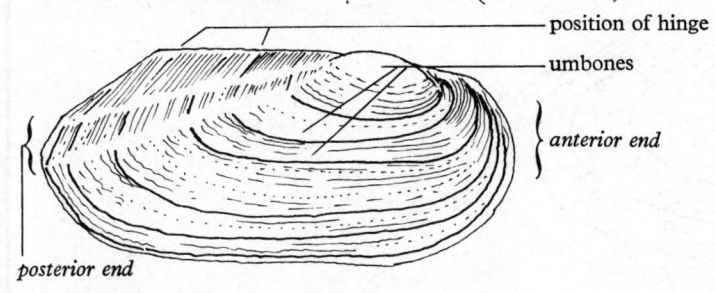

position of hinge
umbones
anterior end
posterior end

Key to families

1 Shell large up to 200 mm long, elongate or broadly ovate with prominent umbones

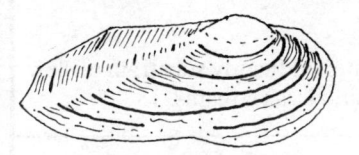

☐ Family: Unionidae
see pages 75–76

Shell usually much smaller, less than 40 mm long

☐ 2

2 Shell triangular in end-on view, striped

☐ Family: Dreissensiidae
see page 77

Shell swollen, rounded, often whitish

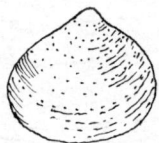

☐ Family: Sphaeriidae
see pages 77–78

Family: Unionidae

× 10

× 10

× 10

Anodonta cygnaea

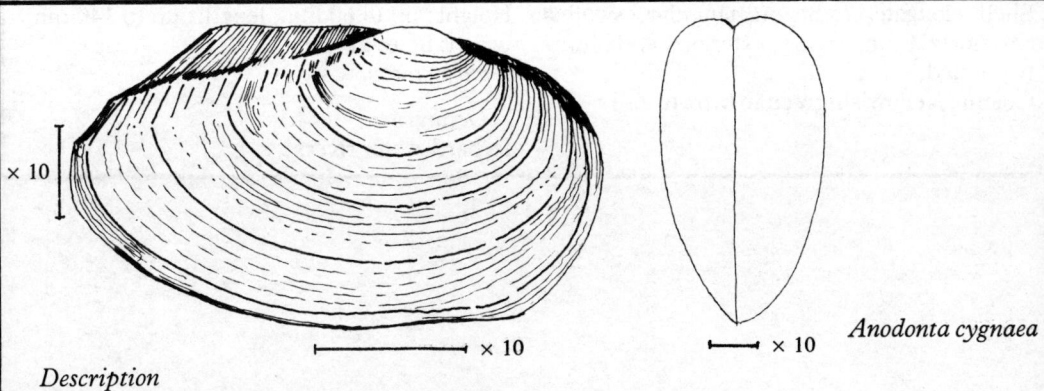

Description
Shell large, elongate-oval with wing-like postero-dorsal processes. Umbones slightly swollen.
Colour: Yellow to yellow-green, but variable
Height: up to 120 mm, length: up to 200 mm, width: up to 60 mm
Anodonta cygnaea

Occurrence
Prefers hard waters in muddy substrata of slow-running rivers

75

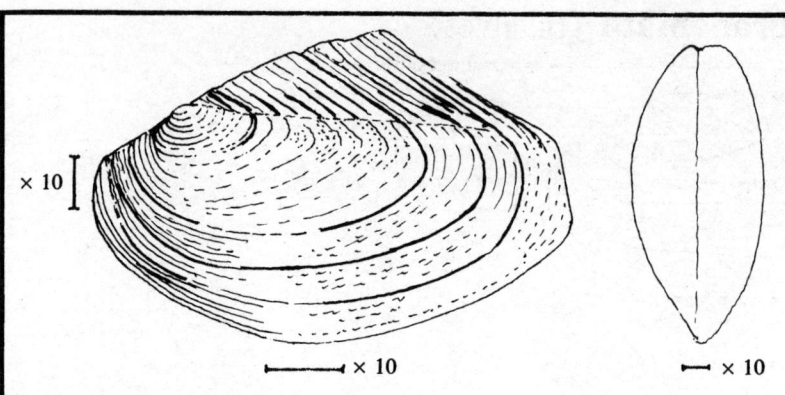

Anodonta anatina

×10

×10

×10

Description
Shell smaller than *Anodonta cygnaea*, outline oval, swollen. Edge of shell more rounded.
Colour: Rather dark green-grey to green-brown
Height: up to 60 mm, length: up to 100 mm, width: up to 30 mm
Anodonta anatina

Occurrence
Prefers hard water and sandy substrata of slow running rivers and large streams

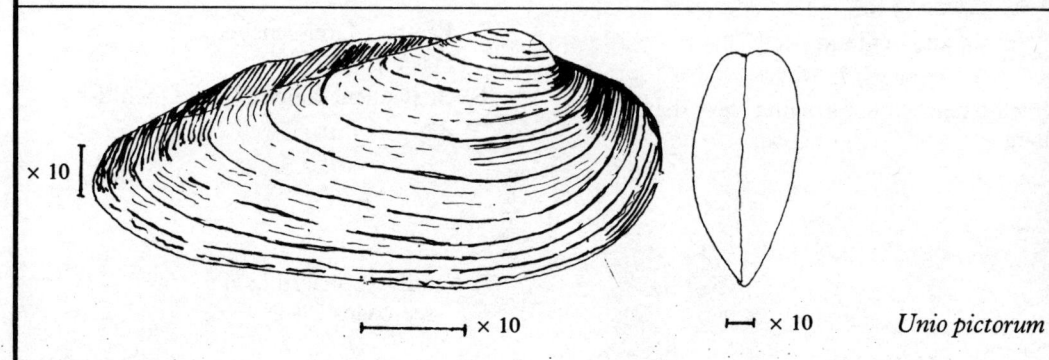

Unio pictorum

×10

×10

Description
Shell elongate, rather rectangular, swollen anteriorly. Compressed posteriorly and bluntly pointed.
Colour: Yellowish-green to brown
Height: up to 60 mm, length: up to 140 mm, width: up to 32 mm
Unio pictorum

Occurrence
In hard-water rivers

Family: Dreissensiidae

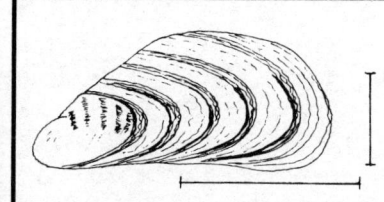

Dreissena polymorpha

Description
Triangular in end-on view. Attached to substratum by threads.
Colour: olive-green, brownish with alternating zig-zag stripes or bands

Height: up to 15 mm, length: up to 30 mm, width: up to 15 mm
Dreissena polymorpha
Occurrence
An invader of fresh water in canals, streams and rivers, etc.

Family: Sphaeriidae

Only three examples of this family are given here. Members of the genus *Pisidium* are particularly difficult to identify and demand considerable expertise. For this reason no attempt has been made in this book to classify to species level.

Key to the genera:

1 Shell usually longer than 10 mm. Umbones anterior to centre. Two long siphons present (observe animal immersed in water)

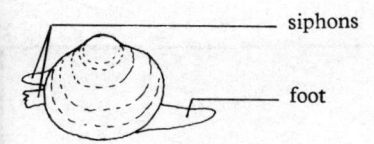

— siphons
— foot

☐ Genus: *Sphaerium*
see pages 77–78

2 Shell usually less than 10 mm long. Umbones posterior to centre. One long siphon present (observe animal immersed in water)

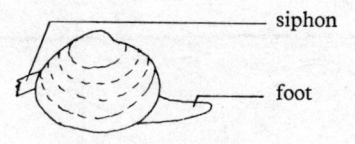

— siphon
— foot

☐ Genus: *Pisidium*
see page 78

Genus: *Sphaerium*

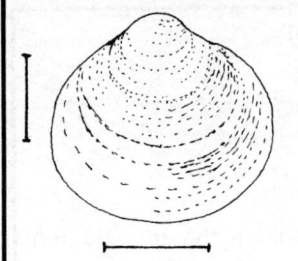

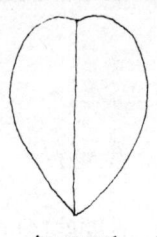

Sphaerium corneum

Description
Shell rounded and swollen, thin and smooth. Umbones small and not prominently raised. Surface with fine irregular striations.
Colour: yellowish, grey or brown with alternating light and dark bands

Height: up to 11 mm, length: up to 14 mm, width: up to 10 mm
Sphaerium corneum

Occurrence
Very common in the mud of clean streams and rivers

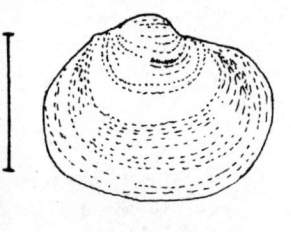

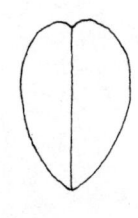

Sphaerium rivicola

Description
A larger species than *Sphaerium corneum*, shell swollen and thick. Outline elliptical. Umbones small and slightly raised, merge into general outline. Concentrically ribbed.
Colour: Yellowish, reddish-brown, olive

Height: up to 18 mm, length: up to 24 mm, widht: up to 14 mm
Sphaerium rivicola

Occurrence
Found in slow-flowing streams and rivers

Genus: *Pisidium*
(Example)

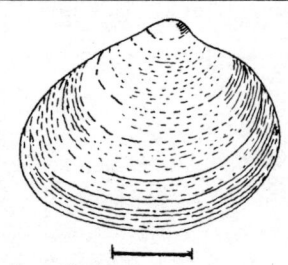

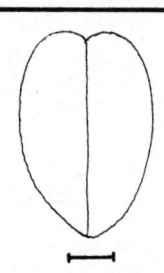

Pisidium amnicum

Description
Shell swollen, thick-walled. Outline oval to rounded-triangular. Umbones broad. Strong irregularly concentric striations.
Colour: Glossy grey-yellow, grey-brown

Height: up to 9 mm, length: up to 11 mm, width: up to 6 mm
Pisidium amnicum

Occurrence
In mud and sand of hard-water streams and rivers

List of references for general reading and more advanced studies

General works:
Mellanby H. (1963) *Animal Life in Freshwater*. Chapman and Hall.
Clegg J. (1965) *Freshwater Life of the British Isles* ('Wayside and Woodland Series'). Warne.
Clegg J. (1956) *The Observer's Book of Pond Life*. Warne. (Keys included).
Engelhardt W. (1964) *The Young Specialist Looks at Pond Life*. Burke.

Ecology:
Hynes N. B. N. (1972) *The Ecology of Running Waters*. Liverpool University Press.
Hynes H. B. N. (1966) *The Biology of Polluted Waters*. Liverpool University Press.
Leadley Brown A. (1971) *Ecology of Freshwater*. Heinemann.

Techniques:
Schwoerbel (1970) *Hydrobiology Freshwater Biology*. Pergamon.

Beetles:
Linssen E. F. (1959) *Beetles of the British Isles* ('Wayside and Woodland Series'). Two volumes. Warne.

Bugs:
Southwood and Leston (1959) *Land and Water Bugs of the British Isles* ('Wayside and Woodland Series'). Warne.

Dragonflies:
Longfield C. (1949) *The Dragonflies of the British Isles* ('Wayside and Woodland Series'). Warne.
Corbet P. S., Longfield C. and Moore N. W. (1960) *Dragonflies* ('The New Naturalist Series'). Collins. (Keys included).

Caddis Larvae:
Hickin Norman E. (1967) *Caddis Larvae*. Hutchinson. (Keys included).

Freshwater Biological Association: Scientific Publications.
5. *A Key to the British Species of Freshwater Cladocera*, D. J. Scourfield and J. P. Harding.
8. *Keys to the British Species of Aquatic Megaloptera and Neuroptera*, D. E. Kimmins.
13. *A Key to the British Fresh- and Brackish-Water Gastropods*, T. T. Macan.
14. *A Key to the British Freshwater Leeches*, K. H. Mann.
16. *A Revised Key to the British Water Bugs (Hemiptera-Heteroptera)*, T. T. Macan.
17. *A Key to the Adults and Nymphs of the British Stoneflies (Plecoptera)*, H. B. N. Hynes.
18. *A Key to the British Freshwater Cyclopid and Calanoid Copepods*, J. P. Harding and W. A. Smith.
19. *A Key to the British Species of Crustacea: Malacostraca occurring in Freshwater*, T. T. Macan, H. B. N. Hynes and W. D. Williams.
20. *A Key to the Nymphs of British Species of Ephemeroptera*, T. T. Macan.
22. *A Guide for the Identification of British Aquatic Oligochaeta*, R. O. Brinkhurst.
23. *A Key to the British Species of Freshwater Triclads*, T. B. Reynoldson.
24. *A Key to the British Species of Simuliidae (Diptera) in the Larval, Pupal and Adult Stages*, Lewis Davies.
26. *A Key to the Larvae, Pupae and Adults of the British Species of Elminthidae*, D. G. Holland.
27. *A Key to the British Freshwater Fishes*, Peter S. Maitland.

These publications may be obtained from: The Librarian, The Ferry House, Ambleside, Cumbria.

Invertebrate animals of streams and rivers

A summary of the various animal groups considered in the guide.

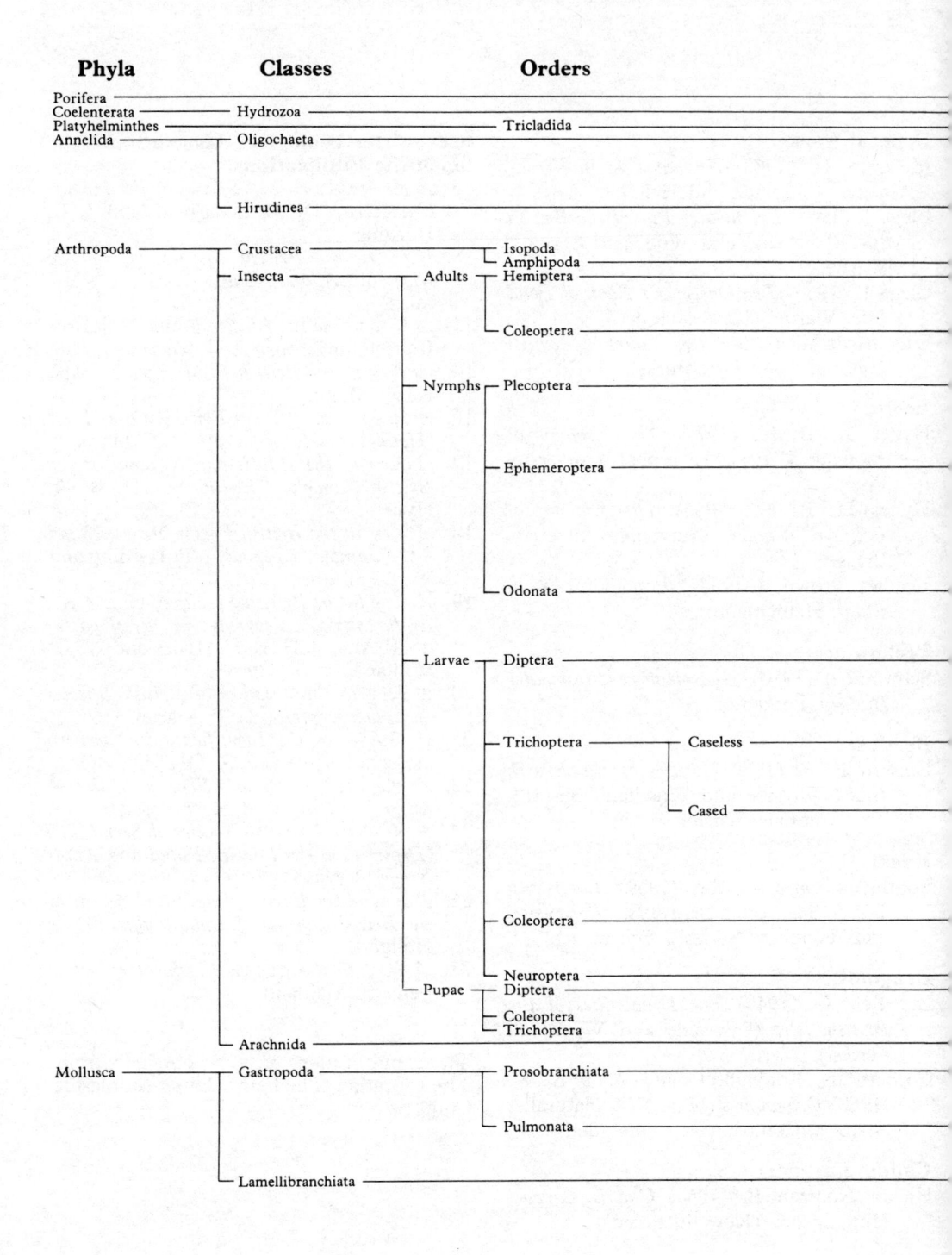

Phyla	Classes	Orders		
Porifera				
Coelenterata	Hydrozoa			
Platyhelminthes		Tricladida		
Annelida	Oligochaeta			
	Hirudinea			
Arthropoda	Crustacea	Isopoda		
		Amphipoda		
	Insecta	Adults — Hemiptera		
		Coleoptera		
		Nymphs — Plecoptera		
		Ephemeroptera		
		Odonata		
		Larvae — Diptera		
		Trichoptera	Caseless	
			Cased	
		Coleoptera		
		Neuroptera		
		Pupae — Diptera		
		Coleoptera		
		Trichoptera		
	Arachnida			
Mollusca	Gastropoda	Prosobranchiata		
		Pulmonata		
	Lamellibranchiata			

80

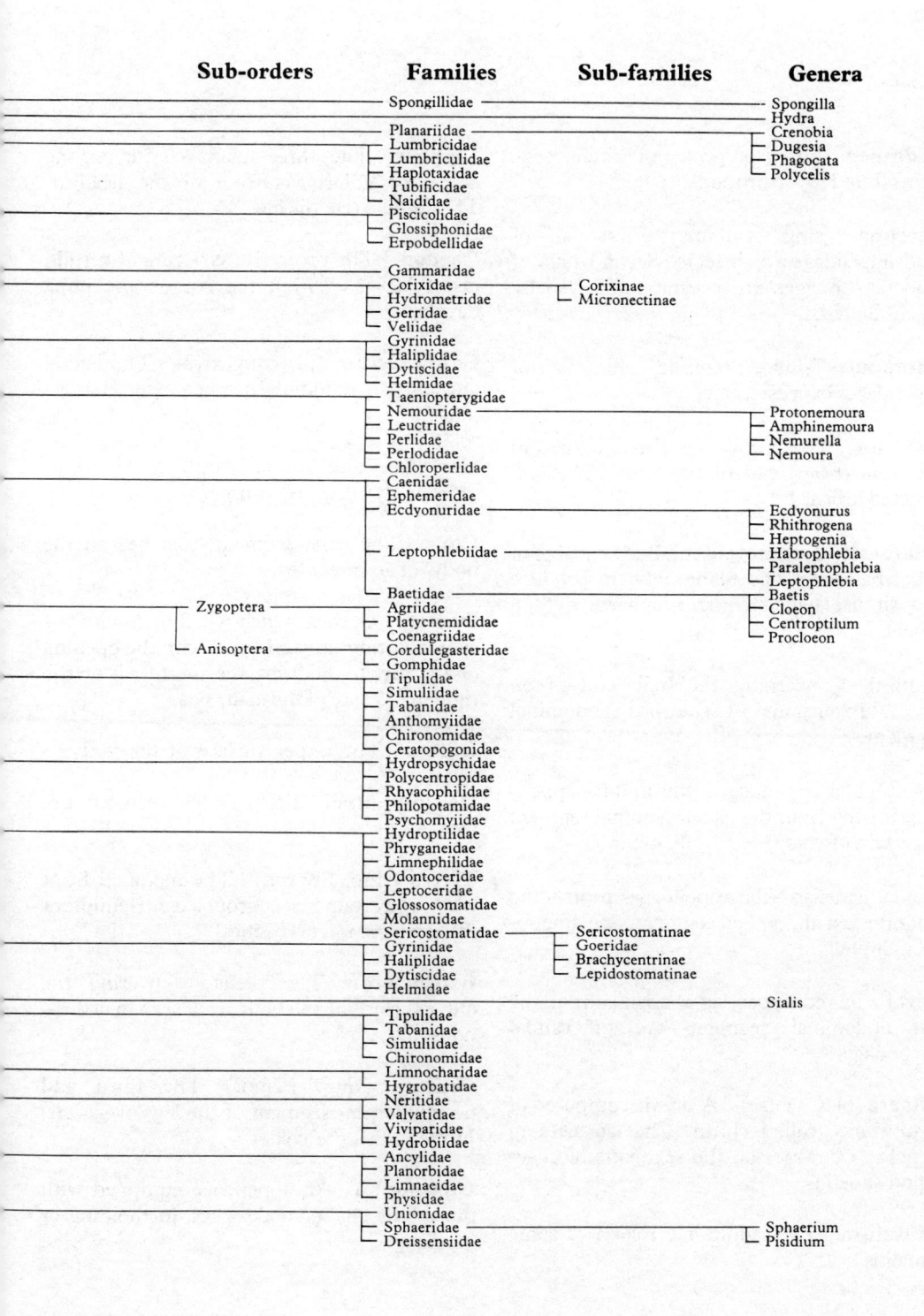

Sub-orders	Families	Sub-families	Genera
	Spongillidae		Spongilla
			Hydra
	Planariidae		Crenobia
	Lumbricidae		Dugesia
	Lumbriculidae		Phagocata
	Haplotaxidae		Polycelis
	Tubificidae		
	Naididae		
	Piscicolidae		
	Glossiphonidae		
	Erpobdellidae		
	Gammaridae		
	Corixidae	Corixinae	
	Hydrometridae	Micronectinae	
	Gerridae		
	Veliidae		
	Gyrinidae		
	Haliplidae		
	Dytiscidae		
	Helmidae		
	Taeniopterygidae		
	Nemouridae		Protonemoura
	Leuctridae		Amphinemoura
	Perlidae		Nemurella
	Perlodidae		Nemoura
	Chloroperlidae		
	Caenidae		
	Ephemeridae		
	Ecdyonuridae		Ecdyonurus
			Rhithrogena
			Heptogenia
	Leptophlebiidae		Habrophlebia
			Paraleptophlebia
			Leptophlebia
	Baetidae		Baetis
Zygoptera	Agriidae		Cloeon
	Platycnemididae		Centroptilum
	Coenagriidae		Procloeon
Anisoptera	Cordulegasteridae		
	Gomphidae		
	Tipulidae		
	Simuliidae		
	Tabanidae		
	Anthomyiidae		
	Chironomidae		
	Ceratopogonidae		
	Hydropsychidae		
	Polycentropidae		
	Rhyacophilidae		
	Philopotamidae		
	Psychomyiidae		
	Hydroptilidae		
	Phryganeidae		
	Limnephilidae		
	Odontoceridae		
	Leptoceridae		
	Glossosomatidae		
	Molannidae		
	Sericostomatidae	Sericostomatinae	
	Gyrinidae	Goeridae	
	Haliplidae	Brachycentrinae	
	Dytiscidae	Lepidostomatinae	
	Helmidae		Sialis
	Tipulidae		
	Tabanidae		
	Simuliidae		
	Chironomidae		
	Limnocharidae		
	Hygrobatidae		
	Neritidae		
	Valvatidae		
	Viviparidae		
	Hydrobiidae		
	Ancylidae		
	Planorbidae		
	Limnaeidae		
	Physidae		
	Unionidae		
	Sphaeridae		Sphaerium
	Dreissensiidae		Pisidium

Glossary

Abdomen Posterior group of similar segments found in Arthropods.

Antennae (sing. Antenna) First pair of head appendages of insects. Second pair of head appendages in crustaceans. Usually much jointed.

Antennules (sing. Antennule) First pair of appendages of crustaceans.

Anterior The part of the body situated at the front (head) end of the body. The end directed forward.

Bilaterally symmetrical Body capable of being halved in one plane only, to produce two similar halves (mirror images of each other).

Caudal Concerning the 'tail' end of an insect. Extensions of the last abdominal segment.

E.g. Caudal appendage—the middle appendage arising from the last abdominal segment of certain insects. See Mayfly nymphs.

Caudal lamellae—the appendages protruding from the last abdominal segment. See dragonfly nymphs.

Cerci Paired appendages arising from the last abdominal segment—tail or caudal appendages.

Chaeta (pl. Chaetae) A bristle composed of a substance called chitin. Chaetae arise in bunches from sacs on the segments of chaetopod annelids.

Clitellum The saddle-like region of some annelids.

Clypeus The three-sided sclerite on the dorsal or anterior surface of the head of Trichoptera (caddis flies.)

Cocoon The protective covering of certain insect larvae. Within the cocoon the pupa develops.

Connexivium (pl: connexiva) The lateral area of each abdominal segment. (see Hemiptera).

Corium The surface covering of wing-cases or elytra. (see Hemiptera)

Coxa The basal segment (that nearest the body) of an insect leg.

Dextral A shell which when held with the apex pointing upwards, and with the opening or aperture to the front, the aperture is on the right-hand side of the main axis.

Dorsal The upper surface of the body.

Dorso-ventral From upper to lower surface.

Elytra (sing. Elytron) The modified front wings of certain insect groups e.g. Hemiptera (bugs), Coleoptera (beetles).

Exoskeleton The skeleton covering the outer surface of the body, well seen in arthropods.

Femora (sing. Femur) The third and usually largest segment of the legs of insects. Jointed with the coxa.

Gnathopod An appendage equipped with pincers or chelipeds designed for holding or grasping.

Hemimetabolous Having an incomplete metamorphosis, without a pupal stage in the life cycle. Three stages in the life cycle: egg—nymph—adult.

Homometabolous Having a complete metamorphosis with a pupal stage in the life cycle. Four stages in the life cycle: egg—larva—pupa—adult.

Hydroid One of the body forms of coelenterates. Possesses a sac-like body and a circlet of tentacles at the free, mouth-end.

Labium The lower lip of insect mouthparts. The paired appendages of one segment fused together.

Lacinia The blade or toothed edge of the maxilla of insect mouthparts.

Lamella A thin plate-like structure.

Larva (pl. Larvae) The young stage of an insect which undergoes complete metamorphosis (four stages in the life-cycle: egg, larva, pupa and adult).

Lateral The side or flank surface of the body.

Mandibles The biting or crushing appendages of the mouthparts of insects and crustaceans.

Maxillary palps One of the pairs of mouthparts of insects and crustaceans.

Mesonotum The dorsal surface of the second thoracic segment of insect thorax.

Metanotum The dorsal surface of the third (last) thoracic segment of an insect thorax.

Nota (sing. Notum) The dorsal surfaces of the thoracic segments, usually prefixed by pro-, meso-, or meta- to indicate the relevant segment.

Nymph The young stage of an insect which undergoes incomplete metamorphosis (three stages in the life-cycle: egg, nymph and adult).

Ocelli (sing. Ocellus) Simple light receptors which are found in many invertebrates.

Operculum The plate covering the opening of the shell in some gastropod molluscs.

Ostia (sing. Ostium) Openings of the surface of sponges through which water enters the body.

Papilla Projection from the surface of the body.

Pereiopods The walking appendages of crustaceans.

Pleopods Paired appendages on the ventral surfaces of the abdominal segments of crustaceans.

Posterior The back or tail end of the body.

Prolegs Stumpy legs on the posterior segments of the larval stages of some insects.

Pronotum The dorsal surface of the first thoracic segment of an insect thorax.

Prothorax The first thoracic segment of the thorax of insects. This segment does not bear wings.

Pupa (pl. Pupae) The resting stage in the life-cycle of insects which undergo complete metamorphosis. During this stage the larval tissue breaks down and the adult is formed.

Radially symmetrical Body capable of being halved in two or more vertical planes to produce two similar halves (mirror images of each other).

Rostrum A bill or beak-shaped part of the anterior region in certain insects.

Sclerotization See sclerotized tissue.

Sclerotized tissue Tissue thickened to form a tough, hard, impermeable layer. Such tissue will appear dark and tough when compared to the surrounding tissue.

Sinistral A shell which when held with the apex pointing upwards, and with the opening or aperture to the front, the aperture is on the left-hand side of the main axis.

Siphon A tube which serves to carry a respiratory and feeding current into the body of lamellibranch molluscs.

Spiracle The external opening to the respiratory system of insects.

Spiracular plate An expanded portion of the last abdominal segment of certain insect larvae on to which the spiracles open.

Sternum The cuticle on the ventral surface of the segments of insects. Often forms a sclerotized plate.

Tarsus (pl. Tarsi) The 'foot' of an insect divided up into several sub-segments.

Telson The last segment of the abdomen of crustaceans.

Tergum (pl. terga) The thickened cuticle on the dorsal surface of the segments of insects.

Tibia (pl. Tibiae) One of the segments of the leg between the femur and the tarsus.

Tubercle A swelling or hump arising from the surface of the body.

Umbilicus A small opening on the surface of the shell above the aperture of certain molluscs.

Umbones Shield-shaped structures on the surface of the shells of lamellibranch molluscs.

Uropods (sing. Uropod) The last pair of abdominal appendages of crustaceans. With the telson, the uropods form the tail fan.

Ventral The lower surface of the body.